Frontiers in Mathematics

Farit G. Avkhadiev
Karl-Joachim Wirths

Schwarz-Pick
Type
Inequalities

Birkhäuser Verlag
Basel · Boston · Berlin

Authors:
Farit G. Avkhadiev
Chebotarev Research Institute
Kazan State University
420008 Kazan
Russia
e-mail: Farit.Avkhadiev@ksu.ru

Karl-Joachim Wirths
Institut für Analysis und Algebra
TU Braunschweig
38106 Braunschweig
Germany
e-mail: kjwirths@tu-bs.de

2000 Mathematical Subject Classification: 30A10, 30C20, 30C55, 30C80, 30D50, 30F45, 51M10

Library of Congress Control Number: 2008943668

Bibliographic information published by Die Deutsche Bibliothek
Die Deutsche Bibliothek lists this publication in the Deutsche Nationalbibliografie;
detailed bibliographic data is available in the Internet at <http://dnb.ddb.de>.

ISBN 978-3-7643-9999-3 Birkhäuser Verlag AG, Basel · Boston · Berlin

© 2009 Birkhäuser Verlag AG
Basel · Boston · Berlin
P.O. Box 133, CH-4010 Basel, Switzerland
Part of Springer Science+Business Media
Cover design: Birgit Blohmann, Zürich, Switzerland
Printed on acid-free paper produced from chlorine-free pulp. TCF ∞
Printed in Germany

ISBN 978-3-7643-9999-3   e-ISBN 978-3-0346-0000-2

9 8 7 6 5 4 3 2 1   www.birkhauser.ch

Dedicated to our families

Contents

Chapter 1

Introduction

The aim of the present book is a unified representation of some recent results in geometric function theory together with a consideration of their historical sources. These results are concerned with functions f, holomorphic or meromorphic in a domain Ω in the extended complex plane $\overline{\mathbb{C}}$. The only additional condition we impose on these functions is the condition that the range $f(\Omega)$ is contained in a given domain $\Pi \subset \overline{\mathbb{C}}$. This fact will be denoted by $f \in A(\Omega, \Pi)$. We shall describe how one may get estimates for the derivatives $|f^{(n)}(z_0)|, n \in \mathbb{N}, f \in A(\Omega, \Pi)$, dependent on the position of z_0 in Ω and $f(z_0)$ in Π.

1.1 Historical remarks

The beginning of this program may be found in the famous article [125] of G. Pick. There, he discusses estimates for the MacLaurin coefficients of functions with positive real part in the unit disc found by C. Carathéodory in [52]. Pick tells his readers that he wants to generalize Carathéodory's estimates such that the special role of the expansion point at the origin is no longer important. For the convenience of our readers we quote this sentence in the original language:

Durch lineare Transformation von z oder, wie man sagen darf, durch kreisgeometrische Verallgemeinerung, kann man die Sonderstellung des Wertes z = 0 wegschaffen, so daß sich Relationen für die Differentialquotienten von w an beliebiger Stelle ergeben.

The first great success of this program was G. Pick's theorem, as it is called by Carathéodory himself, compare [54], vol II, §286–289.

If $z_0 \in \Delta = \{z \mid |z| < 1\}$, $f \in A(\Delta, \Delta)$, and $f(z_0) = w_0$, then the inequality

$$\left| \frac{w_0 - f(z)}{1 - \overline{w_0} f(z)} \right| \le \left| \frac{z_0 - z}{1 - \overline{z_0} z} \right|, \quad z \in \Delta, \tag{1.1}$$

is valid.

This theorem follows immediately from Schwarz's Lemma using holomorphic automorphisms of the unit disc. A direct consequence of Pick's theorem is the inequality

$$|f'(z_0)| \leq \frac{1 - |f(z_0)|^2}{1 - |z_0|^2}, \quad f \in A(\Delta, \Delta), \ z_0 \in \Delta. \tag{1.2}$$

To be more complete about the history of the inequality (1.2) we have to mention that O. Szász (see the footnote in [160], p.308) attributes it to E. Lindelöf [105] and indicates that (we quote once more):

Neuere Beweise dieser Relation gaben Carathéodory und Jensen. Herr Jensen zeigte, wie der Satz aus dem schon von Herrn Landau 1906 bewiesenen Spezialfall: $|c_1| \leq 1 - |c_0|^2$ leicht folgt.

As an oft-quoted proverb says "Success has many fathers, but failure is an orphan". For us, at this place the second part of this proverb hints of unsolved mathematical problems.

It is clear that Schwarz's lemma and its generalizations became widely known in the period of a systematic study of results which are closely connected with the proofs of the Riemann theorem on conformal mappings. By the way, the original version of the lemma may be found in [149], where H.A. Schwarz discussed the Riemann mapping theorem.

Now, it is well known that one may generalize the inequalities (1.1) and (1.2) to $f \in A(\Omega, \Pi), z_0 \in \Omega$, and $f(z_0) = w_0$, where Ω and Π are domains that have at least three boundary points. This generalization is known as the principle of hyperbolic metric (see for instance R. Nevanlinna [120] and G. M. Goluzin [70]).

An important case is presented by simply connected domains. Let Ω and Π be simply connected proper subdomains of $\mathbb{C}$ and let Φ_{Ω, z_0} be the unique conformal map of Δ onto Ω such that $\Phi_{\Omega, z_0}(0) = z_0$ and $\Phi'_{\Omega, z_0}(0) > 0$. The existence of this map is proved by the Riemann mapping theorem. We define $\Phi_{\Pi, f(z_0)}$ analogously. Then the function

$$w = \Phi^{-1}_{\Pi, f(z_0)} \circ f \circ \Phi_{\Omega, z_0} \tag{1.3}$$

belongs to the family $A(\Delta, \Delta)$ and satisfies $w(0) = 0$. Hence, Schwarz's Lemma implies

$$|f'(z_0)| \leq \frac{\Phi'_{\Pi, f(z_0)}(0)}{\Phi'_{\Omega, z_0}(0)},$$

the generalized Schwarz-Pick inequality. The quantities

$$R(z_0, \Omega) := \Phi'_{\Omega, z_0}(0) \quad \text{and} \quad R(f(z_0), \Pi) := \Phi'_{\Pi, f(z_0)}(0)$$

are called the conformal radius of Ω at the point z_0 and of Π at $w_0 = f(z_0)$, respectively, and will be used here to describe the positions of the points z_0 in Ω and $f(z_0)$ in Π.

1.2 On inequalities for higher derivatives

We ask for inequalities of the form

$$\frac{|f^{(n)}(z_0)|}{n!} \leq M_n(z_0, \Omega, \Pi) \frac{R(f(z_0), \Pi)}{(R(z_0, \Omega))^n}, \quad n \in \mathbb{N}, \tag{1.4}$$

where f and z_0 are as above.

Concerning the history of inequalities for higher derivatives, we should mention here that the interest of researchers in geometric function theory was concentrated for a long time after 1920s on the famous Bieberbach conjecture, i.e., the conjecture that

$$\frac{|f^{(n)}(0)|}{n!} \leq n|f'(0)| \tag{1.5}$$

for functions f holomorphic and injective on Δ. This may be seen as a special case of the above inequality, where f maps Δ conformally onto the simply connected domain $f(\Delta)$. Since it seemed very difficult to prove this conjecture, less attention was attracted by the generalized Bieberbach conjecture or Rogosinski conjecture, which in the above formulation means that

$$\frac{|f^{(n)}(0)|}{n!} \leq nR(f(0), \Pi) \tag{1.6}$$

for any simply connected proper subdomain Π of $\mathbb{C}$ and any $f \in A(\Delta, \Pi)$. An equivalent formulation of the Rogosinski conjecture is the following.

Let g be holomorphic and injective on Δ and let there exist a function $w : \Delta \to \Delta$ such that $w(0) = 0$ and $f = g \circ w$. Then (1.6) is valid.

As usual, we will abbreviate this relation between f and g by $f \prec g$ and say that under these circumstances f is subordinated to g.

It was a great surprise when de Branges succeeded in proving not only (1.5) but also (1.6) in 1985. Soon afterwards, it was recognized, especially by Yamashita, that one could use (1.5) to go further steps in Pick's program for functions injective on Δ. In fact, it had been seen earlier by Landau and Jakubowski that the validity of the Bieberbach conjecture would imply sharp bounds for $|f^{(n)}(z_0)|$, $z_0 \in \Delta$, and f injective on Δ. We will give an outline of these results in the present book. Further, we will speak on the ideas of Chua [56], who indicated that (1.5) could be used to derive similar bounds for functions f injective on simply connected domains Ω. In this paper, he considered in addition the case that Ω is convex and published the following conjectures.

Let f be holomorphic and injective on a proper convex subdomain Ω of $\mathbb{C}$. Then for any $z_0 \in \Omega$, and any $n \geq 2$ the inequality

$$\frac{|f^{(n)}(z_0)|}{n!} \leq (n+1)2^{n-2} \frac{|f'(z_0)|}{(R(z_0, \Omega))^{n-1}} \tag{1.7}$$

is valid. If in addition $f(\Omega)$ is convex, then

$$\frac{|f^{(n)}(z_0)|}{n!} \leq 2^{n-1} \frac{|f'(z_0)|}{(R(z_0,\Omega))^{n-1}} \tag{1.8}$$

is valid.

Motivated by the work of Ruscheweyh and Yamashita, who had proved estimates of the type (1.4) and similar ones, the authors of the present book concentrated research in the last years on formulas like (1.4). The results of this research together with known theorems in this direction form the content of this book. Especially, we shall show that Chua's conjectures are true. We will show even more, namely that

$$M_n(z_0,\Omega,\Pi) \leq (n+1)2^{n-2}$$

for $f \in A(\Omega,\Pi), \Omega$ convex and Π simply connected and that

$$M_n(z_0,\Omega,\Pi) \leq 2^{n-1}$$

for $f \in A(\Omega,\Pi)$, Ω and Π convex. These results will be completed by several considerations for special cases of Ω and Π as well as by computations of the dependences of $M_n(z_0,\Omega,\Pi)$ of the variable z_0.

One further implication of the validity of the Bieberbach conjecture is concerned with meromorphic functions univalent on Δ. In 1962, J. Jenkins considered functions f that in addition to the above properties satisfy the condition that the pole of f lies at a point $p \in (0,1)$. He proved that for such functions the Bieberbach conjecture would imply

$$\frac{|f^{(n)}(0)|}{n!} \leq \frac{|f'(0)|}{p^{n-1}} \sum_{k=0}^{n-1} p^{2k}. \tag{1.9}$$

This result motivated us to generalize the above considerations to proper subdomains of $\overline{\mathbb{C}}$. It is natural that here the positions of z_0 in Ω and $f(z_0)$ in Π relative to the point at infinity enter the picture. We use the hyperbolic distances to characterize those items. We will prove results of the type (1.4) using a generalization of the conformal radius. These results are satisfying in the following cases:

1) Ω and Π are simply connected with respect to $\overline{\mathbb{C}}$.

2) Ω is simply connected with respect to $\overline{\mathbb{C}}$ and Π is convex.

3) Ω is convex and $\overline{\mathbb{C}} \setminus \Pi$ is compact and convex.

In the important case that $\Omega = \Delta$ or Ω convex and Π is simply connected with respect to $\overline{\mathbb{C}}$ we can present only partial results.

One of the chapters is devoted to the most difficult case, the case of multiply connected domains. Here, we define appropriate questions and give some answers that in most cases are far from the sharpness we achieved in many of the above mentioned theorems. Hence this chapter is more or less an impetus for further research.

1.3   On methods

What can be said about methods? Essentially, the proofs of Schwarz-Pick type inequalities are based on relationships of certain hyperbolic characteristics of domains with the following results in geometric function theory:

1. Littlewood's ideas on subordinate functions developed by Rogosinski, Goluzin, Clunie, Robertson, and Sheil-Small.

2. Coefficient estimates using Löwner's theory on parametric representation of univalent functions. Especially, Löwner's theorem on inverse coefficients and de Branges' proof of the Bieberbach conjecture.

3. Explicit representation of convex hulls for several families of analytic functions by the Herglotz formula and its generalizations.

Usually, coefficient estimates concern univalent or subordinate functions holomorphic in the unit disk. We prove and use several new versions of the known theorems for subordinate or quasi-subordinate functions which are holomorphic in a neighbourhood of the origin.

Also, to prove Schwarz-Pick type inequalities for higher derivatives one needs certain hyperbolic characteristics of plane domains. For instance, let $Q_n(z, w, \Omega, \Pi)$ be defined as the smallest possible value such that the inequality

$$\frac{|f^{(n)}(z)|}{n!} \leq Q_n(z, w, \Omega, \Pi) \frac{R(w, \Pi)}{R(z, \Omega)^n}$$

holds for all $f \in A(\Omega, \Pi)$, $f(z) = w$, and

$$C_n(\Omega, \Pi) = \sup\{Q_n(z, w, \Omega, \Pi) \mid (z, w) \in \Omega \times \Pi\}.$$

The principle of the hyperbolic metric implies that $Q_1(z, w, \Omega, \Pi) = 1$, and, in turn, that $C_1(\Omega, \Pi) = 1$ for any pair (Ω, Π) equipped with the Poincaré metric.

In the case $n \geq 2$ the quantity Q_n depends on the hyperbolic characteristics

$$R^k(z, \Omega) \frac{\partial^k \log R(z, \Omega)}{\partial z^k} \text{ and } R^k(w, \Pi) \frac{\partial^k \log R(w, \Pi)}{\partial w^k}, \quad 1 \leq k \leq n-1.$$

In particular, $Q_2(z, w, \Omega, \Pi)$ depends on z and w via the quantities $p = |\nabla R(z, \Omega)|$ and $q = |\nabla R(w, \Pi)|$, only. Consequently, explicit estimates of punishing factors $C_n(\Omega, \Pi)$ are closely connected with the behaviour of the above characteristics, and, roughly speaking, estimating of these quantities is related to certain coefficient problems of geometric function theory. Finally, we have to attract the reader's attention to the following fact which deserves to be widely known:

The quantity reciprocal to the conformal or hyperbolic radius is exactly the density of the Poincaré metric. More precisely, the equation

$$\lambda_\Omega(z) := \frac{1}{R(z, \Omega)}, \quad z \in \Omega,$$

defines the density of the hyperbolic metric in the domain Ω with Gaussian curvature $K = -4$.

1.4   Survey of the contents

In this section, we would like to give a short survey of the contents of the present book.

Chapters 2 and 3 have a preparatory character. There we will gather the materials from the work of many researchers in geometric function theory that we need for our results. Therefore we do not give all proofs in detail in these two chapters and we shortened and simplified old proofs. During these efforts we found "new looks through old holes" now and then.

Chapter 2 is dedicated to the most famous coefficient theorems of the last century, namely the Bieberbach conjecture and the conjecture for the coefficients of functions with bounded boundary rotation.

The third chapter discusses the second theme that is important for our generalizations of the Schwarz-Pick lemma; the Poincaré metric, its historical background, and well-known theorems concerning this metric in different circumstances. Among them are the famous theorems of Landau, Teichmüller, Beardon and Pommerenke.

Most of the material presented in the following chapters has been developed by us in the last ten years, some of it has been published, other results appear in this book for the first time.

Chapter 4 contains the most prominent members of the family of punishing factors, those for pairs of simply connected domains and those for pairs of convex domains. Moreover, these two are in some sense the most beautiful, namely 4^{n-1} and 2^{n-1}. We discuss the work of Ruscheweyh, Chua, and Yamashita on these questions. Further, we present the complete solution for the case $n = 2$ for any pair of domains.

The fifth chapter is devoted to more special results. The most prominent of them may be the proof of a far reaching generalization of Chua's conjecture. We add the determination of punishing factors for pairs of simply connected domains in the extended plane, and we prove sharp lower bounds in some general circumstances.

Chapter 6 is concerned with some generalizations to multiply connected domains for arbitrary $n > 2$ and it has at some places a tentative character, since we are not sure that we have found the "best" way of generalization.

In the last two chapters, we present some material in the neighbourhood of our results. This is meant as the basis for further research on the many questions that are natural to pose here.

At this point, we want to thank the Deutsche Forschungsgemeinschaft for continued support of our research. Their many grants for F. G. Avkhadiev enabled us to do all the scientific work presented in this book.

Chapter 2

Basic coefficient inequalities

There are many books that systematically present coefficient problems in geometric function theory. We refer the reader to the excellent monographs by Goluzin [70], Goodman [73], Hayman [78], Pommerenke [128], and Duren [60]. In this chapter we only mention a few classical results on coefficients which are closely connected with the topic of this book. Also, we give several new facts with short proofs that have until now been presented only in original papers.

2.1 Subordinate functions

Let the functions Φ and Ψ be meromorphic in the unit disc Δ. We will say that Φ is subordinate to Ψ and write $\Phi \prec \Psi$ or

$$\Phi(z) \prec \Psi(z),$$

whenever there exists a function ω holomorphic in Δ with properties $\omega(0) = 0$, $|\omega(z)| < 1$, and such that

$$\Phi(z) = \Psi(\omega(z)).$$

Clearly, $|\omega(z)| \leq |z|$ in Δ by the Schwarz lemma.

In [106], Littlewood proved the following result.

Theorem 2.1. *Let Φ and Ψ be holomorphic in the unit disc Δ, and let $\Phi(0) = \Psi(0) = 0$. If $\Phi(z) \prec \Psi(z)$ and $p \in (0, \infty)$, then for any $r \in (0, 1)$,*

$$\int_0^{2\pi} |\Phi(re^{i\theta})|^p d\theta \leq \int_0^{2\pi} |\Psi(re^{i\theta})|^p d\theta. \tag{2.1}$$

We reproduce a short proof of Theorem 2.1 for the case $p \in \mathbb{N}$, only.

Proof of Theorem 2.1 *in the case* $p \in \mathbb{N}$. By Poisson's formula,

$$\Psi^p(z) = \frac{1}{2\pi} \int_0^{2\pi} \Psi^p(\zeta) \operatorname{Re}\left(\frac{\zeta + z}{\zeta - z}\right) d\theta, \quad \zeta = re^{i\theta}, |z| < r < 1.$$

For $|\omega(z)| \leq |z| < r$ we can write

$$|\Phi(z)|^p = |\Psi(\omega(z))|^p \leq \frac{1}{2\pi} \int_0^{2\pi} |\Psi(\zeta)|^p \operatorname{Re}\left(\frac{\zeta + \omega(z)}{\zeta - \omega(z)}\right) d\theta, \quad \zeta = re^{i\theta}.$$

Integrating the latter inequality over the circle $\{z = \rho e^{it} \mid 0 \leq t \leq 2\pi\}$, $\rho \in (0, r)$, using

$$\frac{1}{2\pi} \int_0^{2\pi} \operatorname{Re}\left(\frac{\zeta + \omega(\rho e^{it})}{\zeta - \omega(\rho e^{it})}\right) dt = \operatorname{Re}\left(\frac{\zeta + \omega(0)}{\zeta - \omega(0)}\right) = 1,$$

one gets

$$\int_0^{2\pi} |\Phi(\rho e^{it})|^p dt \leq \int_0^{2\pi} |\Psi(re^{i\theta})|^p d\theta, \quad 0 < \rho < r < 1.$$

Letting $\rho \to r$ gives inequality (2.1) in the case $p \in \mathbb{N}$. $\qquad\qquad\qquad\square$

Remark 2.2. Clearly, a direct use of Theorem 2.1 with $p = 2$ and Parseval's formula for the holomorphic functions

$$\Phi(z) = \sum_{n=0}^{\infty} A_n z^n \quad and \quad \Psi(z) = \sum_{n=0}^{\infty} B_n z^n, \quad |z| < 1,$$

leads to the inequality

$$\sum_{n=0}^{\infty} |A_n|^2 \leq \sum_{n=0}^{\infty} |B_n|^2.$$

Using Theorem 2.1 with $p = 2$ in an original way, Rogosinski proved the following assertion on coefficients of these two functions.

Theorem 2.3 (Rogosinski, [139]). *Let Φ and Ψ be holomorphic in the unit disc Δ. If $\Phi(z) \prec \Psi(z)$, then for all $n \in \mathbb{N} \cup \{0\}$,*

$$\sum_{k=0}^{n} |A_k|^2 \leq \sum_{k=0}^{n} |B_k|^2.$$

Simple counterexamples show that the inequality $|A_n| \leq |B_n|$ for $n \geq 2$ does not hold, for instance, for the functions $\Phi(z) = z^n$ and $\Psi(z) = z$. Also, the assertion of Theorem 2.3 with $p \neq 2$, i.e., the inequality

$$\sum_{k=0}^{n} |A_k|^p \leq \sum_{k=0}^{n} |B_k|^p$$

is not true in general (see [139]).

In [137], Robertson generalized Theorem 2.3 to the case when the functions Φ and Ψ are holomorphic in the unit disc Δ and

$$\Phi(z) = \varphi(z)\Psi(\omega(z)), \quad |z| < 1,$$

where $|\varphi(z)| \leq 1$ for $z \in \Delta$ and ω is as above. In this case Φ is said to be **quasi-subordinate** to the function Ψ (see also [57] and [128]).

Robertson's theorem can be generalized to meromorphic functions or, more generally, to functions F and G that are holomorphic only in a neighbourhood of the origin (see [19], [30] and [20]).

Theorem 2.4 (see [20]). *Let the functions F and G be holomorphic in a neighbourhood of the origin, where they have expansions*

$$F(z) = \sum_{n=0}^{\infty} A_n z^n \quad and \quad G(z) = \sum_{n=0}^{\infty} B_n z^n.$$

If there exist two functions φ and ω holomorphic in the unit disc Δ with $|\varphi(z)| \leq 1$ and $|\omega(z)| \leq |z|$ for $z \in \Delta$ such that the identity

$$F(z) = \varphi(z)G(\omega(z)) \tag{2.2}$$

is satisfied in a neighbourhood of the origin, then for all $n \in \mathbb{N} \cup \{0\}$ the following inequalities are valid:

$$\sum_{k=0}^{n} |A_k|^2 \leq \sum_{k=0}^{n} |B_k|^2. \tag{2.3}$$

Proof. The proof differs from the classical one only in some details.
We fix $n \in \mathbb{N} \cup \{0\}$. From (2.2) we conclude that, in a neighbourhood of the origin,

$$\sum_{k=n+1}^{\infty} A_k z^k - \varphi(z) \sum_{k=n+1}^{\infty} B_k\, \omega(z)^k = \varphi(z) \sum_{k=0}^{n} B_k\, \omega(z)^k - \sum_{k=0}^{n} A_k z^k.$$

If we expand the difference on the left side of this equation in a Taylor series in a neighbourhood of the origin, we see that the coefficients of $z^k, 0 \leq k \leq n$, in this series are zero. Therefore, we get, if we denote this function by H, an expansion

$$H(z) = \sum_{k=n+1}^{\infty} C_k z^k$$

in a neighbourhood of the origin. On the other hand, it is immediately clear that the difference on the right side is holomorphic in the whole unit disc. This implies that the identity

$$\sum_{k=0}^{n} A_k z^k + \sum_{k=n+1}^{\infty} C_k z^k = \varphi(z) \sum_{k=0}^{n} B_k\, \omega(z)^k$$

is valid for $|z| < 1$, too.

Now, we observe that the Littlewood theorem is true for quasi-subordinate holomorphic functions, too. Consequently, we may proceed as in the classical proof of Theorem 2.3 to get

$$\sum_{k=0}^{n} |A_k|^2 \leq \sum_{k=0}^{n} |A_k|^2 + \sum_{k=n+1}^{\infty} |C_k|^2 \leq \sum_{k=0}^{n} |B_k|^2$$

by use of Parseval's formula in Theorem 2.1 with $p = 2$ for the quasi-subordinate holomorphic functions

$$\Phi(z) = \sum_{k=0}^{n} A_k z^k + \sum_{k=n+1}^{\infty} C_k z^k, \quad \Psi(z) = \sum_{k=0}^{n} B_k \, z^k, \quad z \in \Delta.$$

This completes the proof of Theorem 2.4. $\square$

Following the idea of Goluzin (see [70] and compare also [57] and [137], Theorem 6.3) one may obtain a generalization of Theorem 2.4 as follows. Consider inequality (2.3) for $n = 0, 1, \ldots, m$, where $m \in \mathbb{N}$. Let $\lambda_0 \geq \lambda_1 \geq \cdots \geq \lambda_m \geq \lambda_{m+1} = 0$. We multiply the inequality (2.3) for $n = j$ by $\lambda_j - \lambda_{j+1} \geq 0$, $j = 0, 1, \ldots, m$. Summing up the results over j one easily has

$$\sum_{j=0}^{m} \lambda_j |A_j|^2 \leq \sum_{j=0}^{m} \lambda_j |B_j|^2.$$

Since m is arbitrary, this gives the following theorem, Goluzin's version of the inequalities (2.3).

Theorem 2.5. *Let $n \in \mathbb{N} \cup \{0\}$ and let F and G be as in Theorem 2.4. If*

$$\lambda_0 \geq \lambda_1 \geq \ldots \lambda_n \geq 0,$$

then the following inequality is valid:

$$\sum_{k=0}^{n} \lambda_k |A_k|^2 \leq \sum_{k=0}^{n} \lambda_k |B_k|^2. \tag{2.4}$$

The following assertion generalizes a known idea due to Clunie [57].

Corollary 2.6. *Let $n \in \mathbb{N} \cup \{0\}$ and let f and g be functions meromorphic in the unit disc Δ with expansions of the form*

$$f(z) = \sum_{k=0}^{\infty} A_k z^{k-m} \quad and \quad g(z) = \sum_{k=0}^{\infty} B_k z^{k-m}$$

in a neighbourhood of the origin with some $m \in \mathbb{N}$. If $\lambda_0 \geq \lambda_1 \geq \ldots \lambda_n \geq 0$ and $|f(z)| \leq |g(z)|$ in Δ, then the inequality (2.4) is valid.

To obtain Corollary 2.6 it is sufficient to apply Theorem 2.5 to functions F and G defined by

$$F(z) = z^m f(z),\ G(z) = z^m g(z),\ \omega(z) = z,\ \varphi(z) = f(z)/g(z).$$

2.2   Bieberbach's conjecture by de Branges

Consider the normal family S of all functions f that are holomorphic and univalent in Δ and have a Taylor expansion of the form

$$f(z) = z + \sum_{n=2}^{\infty} a_n z^n, \quad |z| < 1.$$

Before de Branges' proof of the Bieberbach conjecture in [43] via the Milin conjecture, the following seven conjectures in their full generality were open problems.

1. **Bieberbach Conjecture (1916).** ([40]) *For any $f \in S$ the inequality $|a_n| \le n$ holds for all $n \ge 2$. The equality occurs if and only if $f(z)$ is the Koebe function $k_d(z) = z(1 + dz)^{-2}$, $|d| = 1$.*

2. **Littlewood Conjecture (1925).** ([106]) *If $f \in S$ and $f(z) \ne w$ for any $z \in \Delta$, then $|a_n| \le 4|w|n$ holds for all $n \ge 2$.*

3. **Robertson Conjecture (1936).** ([136]) *For any odd function $h(z) = z + c_3 z^3 + c_5 z^5 + \cdots$ in S, the inequality*

$$1 + |c_3|^2 + \cdots + |c_{2n-1}|^2 \le n,$$

   is true for all $n \ge 2$.

4. **Rogosinski (Generalized Bieberbach) Conjecture (1943).** ([139]) *Let $g(z) = b_1 z + \cdots + b_n z^n + \cdots$ be a holomorphic function in Δ. If $g(\Delta) \subset f(\Delta)$ and $f \in S$, then the inequality $|b_n| \le n$ holds for all $n \ge 2$.*

5. **Asymptotic Bieberbach Conjecture (1955), connected with Hayman's regularity theorem** (see [78]). *If*

$$A_n = \max_{f \in S} |a_n|,$$

   then

$$\lim_{n \to \infty} \frac{A_n}{n} = 1.$$

6. **Milin Conjecture (1967).** ([116]) *For any $f \in S$, let γ_n be defined by*

$$\log \frac{f(z)}{z} = 2 \sum_{n=1}^{\infty} \gamma_n z^n.$$

Then the inequality

$$\sum_{m=1}^{n}\sum_{k=1}^{m}\left(k|\gamma_k|^2 - \frac{1}{k}\right) \leq 0$$

holds for all $n \geq 1$.

7. **Sheil-Small Conjecture (1973).** ([152]) *For any $f \in S$ and any polynomial $P(z) = b_0 + b_1 z + \cdots + b_n z^n$ the convolution (= Hadamard product) defined by $(P * f)(z) = a_1 z + a_2 b_2 z^2 + \cdots + a_n b_n z^n$ satisfies the inequality*

$$\max_{|z|\leq 1}|(P * f)(z)| \leq n \max_{|z|\leq 1}|P(z)|,$$

for all $n \geq 2$.

We refer the reader to the nice book [71] concerning details of the logical non-trivial relationship between these seven conjectures. In general, there are the following implications:

$$\begin{aligned}
&\quad\textbf{Milin Conjecture} &\Longrightarrow\quad &\textbf{Robertson Conjecture}\\
\Longrightarrow\ &\textbf{Sheil-Small Conjecture} &\Longrightarrow\quad &\textbf{Rogosinski Conjecture}\\
\Longrightarrow\ &\textbf{Bieberbach Conjecture} & &\\
\Longrightarrow\ &\textbf{Asymptotic Bieberbach Conjecture} &\Longrightarrow\quad &\textbf{Littlewood Conjecture.}
\end{aligned}$$

Thus, in [43] de Branges settled all these conjectures by proving the Milin conjecture although he considered only the implications

Milin Conjecture $\Rightarrow$ Robertson Conjecture $\Rightarrow$ Bieberbach Conjecture.

In connection with these seven conjectures, in the next section we shall examine with proof two facts which deserve to be better known. The first one concerns a conjecture by Goodman (see [72] and [73]), which says that the Taylor coefficients of the function $f_p(z) = z + \sum_{n=2}^{\infty} a_n(p)z^n$, $|z| < p$, satisfy the sharp inequality

$$|a_n(p)| \leq \left(1 + p^2 + \cdots + p^{2n-2}\right)/p^{n-1},$$

whenever the function f is meromorphic and univalent in Δ and f has a simple pole at a point $pe^{it} \in \Delta$, $0 < p < 1$. Letting $p \to 1$ this implies the inequality $|a_n| \leq n$ conjectured by Bieberbach. In fact, the Bieberbach conjecture is equivalent to **the Goodman conjecture (1956)** by an elegant proof of Jenkins [86].

Secondly, we shall consider the Sheil-Small conjecture in its full generality (see [152]), which deals with functions subordinate to $f \in S$. This little nuance becomes important in applications to the Schwarz-Pick type inequalities. Moreover, this consideration will contain the whole path from the Robertson conjecture to the Rogosinski conjecture as a special case.

For the convenience of the reader who is not familiar with these inequalities, we will explain without technical details the relationships between the different conjectures. The easiest step is the implication

Robertson conjecture $\Longrightarrow$ Bieberbach conjecture.

For $f \in S$ let $\tilde{h}(z) = \sqrt{f(z^2)/z^2}$. Then the odd function

$$h(z) = z\tilde{h}(z) = z + \sum_{n=2}^{\infty} c_{2n-1} z^{2n-1}$$

belongs to the family S as well. If we take into account that $\tilde{h}$ is an even function we see that $s(z) = \tilde{h}(\sqrt{z})$ is holomorphic in Δ and that $f(z) = z(s(z))^2$. This identity implies that, for $n \geq 2$,

$$a_n = \sum_{k=0}^{n-1} c_{2k+1} c_{2(n-k)-1},$$

where $c_1 = 1$.

According to the Cauchy inequality, this yields

$$|a_n| \leq \sum_{k=0}^{n-1} |c_{2k+1}|^2.$$

Hence, the above implication is obvious. In fact, in the present book, we will need a more general implication from the truth of the Robertson conjecture, namely, the theorem of Sheil-Small, see below.

The implication

Milin conjecture $\Longrightarrow$ Robertson conjecture

follows from a theorem that is concerned with the exponentiation of holomorphic functions, the so-called **Second Lebedev-Milin Inequality** (see [99] and [116]). Let

$$\phi(z) = \sum_{k=1}^{\infty} \alpha_k z^k$$

be holomorphic in a neighbourhood of the origin and

$$e^{\phi(z)} = \sum_{k=0}^{\infty} \beta_k z^k.$$

Then for $n \geq 2$ the inequalities

$$\frac{1}{n} \sum_{k=0}^{n-1} |\beta_k|^2 \leq \exp\left(\frac{1}{n} \sum_{m=1}^{n-1} \sum_{k=1}^{m} \left(k|\alpha_k|^2 - \frac{1}{k} \right) \right)$$

are valid.

If one applies this theorem to the function

$$\phi(z) = \log(s(z)) = \sum_{k=1}^{\infty} \gamma_k z^k,$$

one sees that to prove the Robertson conjecture it is sufficient to prove that for $n \in \mathbb{N}$ the inequalities

$$\sum_{m=1}^{n} \sum_{k=1}^{m} \left(k|\gamma_k|^2 - \frac{1}{k} \right) = \sum_{k=1}^{n} \left(k|\gamma_k|^2 - \frac{1}{k} \right)(n-k+1) \leq 0$$

are valid.

It has become customary to formulate this inequality in terms of the so-called *logarithmic coefficients* of f. Since

$$\log(f(z)/z) = 2\log(s(z)) = \sum_{k=1}^{\infty} 2\gamma_k z^k,$$

many people know this conjecture in the form

$$\sum_{k=1}^{n} \left(k|2\gamma_k|^2 - \frac{4}{k} \right)(n-k+1) \leq 0. \tag{2.5}$$

From this formulation de Branges found his way to prove the Milin conjecture (see [43], [64], and [71]).

The first item in this proof is the Löwner theory assuring firstly that, in problems like the above, it is sufficient to consider univalent functions that map the unit disc onto the complex plane minus a slit. The second ingredient from Löwner's theory is the fact that for any such function f there exists a chain of functions

$$f(z,t) = e^t z + \sum_{n=2}^{\infty} a_n(t) z^n, \quad z \in \Delta, t \in [0, \infty),$$

such that $f(z,0) = f(z)$ and the Löwner differential equation

$$\frac{\partial f(z,t)}{\partial t} = \frac{1 + \kappa(t)}{1 - \kappa(t)} \frac{z \partial f(z,t)}{\partial z}$$

is satisfied, where $|\kappa(t)| = 1$ and κ continuous on $[0, \infty)$.

Naturally, this differential equation results in differential equations for the logarithmic coefficients $c_n(t)$ defined by

$$\log \left(\frac{f(z,t)}{e^t z} \right) = \sum_{n=1}^{\infty} c_n(t) z^n$$

with $c_n(0) = 2\gamma_n$. The genial idea of de Branges was to look at (2.5) as an initial value problem. He constructed the so-called *special function system of de Branges* $\tau_{n,k}(t)$, $1 \leq k \leq n$, $n \in \mathbb{N}$, $t \in [0, \infty)$, such that $\tau_{n,k}(0) = n - k + 1$ considering the functions

$$\varphi_n(t) = \sum_{k=1}^{n} \left(k|c_k(t)|^2 - \frac{4}{k} \right) \tau_{n,k}(t).$$

Using the Löwner differential equation he showed that for his system of functions
the identity

$$\varphi_n'(t) = -\sum_{k=1}^{n} |b_{k-1}(t) + b_k(t) + 2|^2 \frac{\tau_{n,k}'(t)}{k}$$

is valid, where

$$b_0(t) = 0, b_k(t) = \sum_{j=1}^{k} jc_j(t)\kappa(t)^{-j}, \quad k \in \mathbb{N}.$$

The ingredient of the theory of special functions in de Branges' proof was the
proof that $\tau_{n,k}'(t) \leq 0$ and $\lim_{t\to\infty} \tau_{n,k}(t) = 0$.

Hence, $\lim_{t\to\infty} \varphi_n(t) = 0$ and $\varphi_n'(t) \geq 0$. This in turn implies $\varphi_n(0) \leq 0$
which is equivalent to (2.5).

2.3   Theorems of Jenkins and Sheil-Small

Let $p = \in (0,1]$. We need the expansion

$$\kappa_p(z) = \frac{z}{(1-pz)\left(1-\frac{z}{p}\right)} = z + \sum_{n=2}^{\infty} c_n(p)z^n, \quad |z| < p. \tag{2.6}$$

It is known that (see [86])

$$c_n(p) = \frac{\sum_{j=0}^{n-1} p^{2j}}{p^{n-1}} = \frac{p^n - \frac{1}{p^n}}{p - \frac{1}{p}}. \tag{2.7}$$

In 1962, when the proof of the Bieberbach conjecture was a far-off dream,
Jenkins proved the following theorem.

Theorem 2.7 (Jenkins [86]). *Let f_p be a function meromorphic and univalent in
the unit disc Δ with a simple pole at the point pe^{it}, $t \in \mathbb{R}$, and $0 < p < 1$, and let
f_p have the expansion*

$$f_p(z) = z + \sum_{n=2}^{\infty} a_n(p)z^n \tag{2.8}$$

*in a neighbourhood of the origin. If the Bieberbach conjecture is true for all coef-
ficients of schlicht functions, then $|a_n(p)| \leq c_n(p)$ for any $n \geq 2$. Equality occurs
for the function $e^{it}\kappa_p(e^{-it}z)$.*

Proof. Without loss of generality we suppose that $t = 0$. Let $\Delta(p)$ be the domain
obtained from the unit disc by deleting the segment $[p, 1]$. Consider the class $S(p)$
of all functions holomorphic and univalent in $\Delta(p)$ with expansion of the form

(2.8) in a neighbourhood of the origin. It is obvious that $S(1) = S$. To find a conformal map of $\Delta(p)$ onto Δ we proceed as follows. It is clear that the function

$$\tilde{\kappa}_p(z) = \frac{(1+p)^2}{4p} \kappa_p(z)$$

maps $\Delta(p)$ conformally onto $\overline{\mathbb{C}} \setminus (-\infty, -1/4]$. Now, we use the inverse K_{-1} of the Koebe function $k_{-1}(z) = z/(1-z)^2$ to map this domain onto Δ. Therefore, the function $\varphi = K_{-1} \circ \tilde{\kappa}_p$ has the desired property. If we consider the expansion

$$\varphi(z) = \frac{(1+p)^2}{4p} z + \sum_{n=2}^{\infty} c_n z^n,$$

it is important to recognize that $c_n > 0$ for all n. To see this, we compute φ explicitly by the above procedure to get

$$\varphi(z) = 1 + \frac{2p(1 - z/p)(1 - zp)}{(1+p)^2 z} - \frac{2p(1 + z)(1 - z/p)^{1/2}(1 - zp)^{1/2}}{(1+p)^2 z}.$$

It is clear that $c_n, n \geq 2$, is the sum of two consecutive coefficients of the function $(1 - z/p)^{1/2}(1 - zp)^{1/2}$ multiplied by the factor $-2p/(1+p)^2$. The fact that these coefficients themselves are negative for $p \in (0, 1)$ is easily seen using the generalized binomial expansion and the Cauchy product for the product $(1 - z/p)^{1/2}(1 - zp)^{1/2}$.

The inverse to the function φ is frequently used in extremal problems for bounded functions holomorphic and univalent in the unit disc (see for instance [133]).

Clearly, any function $f_p \in S(p)$ admits a representation of the form

$$f_p(z) = \frac{4p}{(1+p)^2} f(\varphi(z)), \quad z \in \Delta(p),$$

where

$$f(z) = z + \sum_{n=2}^{\infty} a_n z^n, \quad z \in \Delta,$$

is a function in S. The function $f_p(z) = 4p(1 + p)^{-2} f(\varphi(z))$ has the expansion of the form (2.8) with

$$a_n(p) = \lambda_1(n) + \sum_{j=2}^{n} \lambda_j(n) a_j,$$

where $\lambda_j(n)$ are polynomials in $c_m = \varphi^{(m)}(0)/m!$, $m = 2, \ldots, n$, with non-negative coefficients, and thus they are themselves non-negative. In particular, if $a_j = j$, i.e. if $f(z) = \kappa_1(z)$, then $a_n(p) = c_n(p)$.

The validity of the Bieberbach conjecture implies that

$$|a_n(p)| \le \lambda_1(n) + \sum_{j=2}^{n} \lambda_j(n)|a_j|$$

$$\le \lambda_1(n) + \sum_{j=2}^{n} j\lambda_j(n) = c_n(p).$$

This completes the proof of Theorem 2.7.   □

Remark 2.8. As is shown in [18], the domain of variablity of coefficients $a_n(p)$ is only a proper subset of the set $\{w \mid |w| \le c_n(p)\}$, if the pole at p lies near to the origin.

The formulation of the Sheil-Small result in its full generality that we have in mind and which can be found, at least implicitly, in [152] and [71] is the following.

Theorem 2.9 (Sheil-Small [152] (1973)). *Let g be subordinated to a function $f \in S$ and P a polynomial of degree less than or equal to n. If the Robertson conjecture is true, then*

$$\max_{|z|\le 1} |(P * g)(z)| \le n \max_{|z|=1} |P(z)| =: n\, M(P). \tag{2.9}$$

First we consider the following lemma.

Lemma 2.10 (Sheil-Small [152]). *Let*

$$U(z) = \sum_{k=0}^{\infty} u_k z^k \quad and \quad V(z) = \sum_{k=0}^{\infty} v_k z^k$$

be holomorphic in Δ and $w_i : \Delta \to \overline{\Delta}, i = 1, 2, 3$, be holomorphic in Δ. If P is a polynomial of degree less than or equal to n and

$$\tilde{h}(z) = z w_1(z) U(z w_2(z)) V(z w_3(z)), \quad z \in \Delta,$$

then we have

$$|(P * \tilde{h})(re^{i\theta})| \le r \max_{|z|=1} |P(z)| \left(\sum_{k=0}^{n-1} |u_k|^2 r^{2k} \right)^{1/2} \left(\sum_{k=0}^{n-1} |v_k|^2 r^{2k} \right)^{1/2} \tag{2.10}$$

for $r \in (0,1)$, $\theta \in [0, 2\pi]$.

Proof. Let the abbreviations be as above. We use the representations

$$(P * \tilde{h})(re^{i\theta}) = \frac{1}{2\pi} \int_0^{2\pi} P(e^{i(\theta - \varphi)}) \tilde{h}(re^{i(\theta + \varphi)})\, d\varphi.$$

and

$$(P * \tilde{h})(z) = P(z) * \left(zw_1(z) \sum_{k=0}^{n-1} u_k(zw_2(z))^k \sum_{k=0}^{n-1} v_k(zw_3(z))^k \right).$$

From this, we obtain

$$|(P * \tilde{h})(re^{i\theta})|$$

$$\leq r\,M(P)\frac{1}{2\pi}\int_0^{2\pi}|w_1(re^{i\varphi})|\left|\sum_{k=0}^{n-1}u_k(re^{i\varphi}w_2(re^{i\varphi}))^k\right|\left|\sum_{k=0}^{n-1}v_k(re^{i\varphi}w_3(re^{i\varphi}))^k\right|\,d\varphi$$

$$\leq r\,M(P)\left(\frac{1}{2\pi}\int_0^{2\pi}\left|\sum_{k=0}^{n-1}u_k(re^{i\varphi}w_2(re^{i\varphi}))^k\right|^2\,d\varphi\right)^{1/2}$$

$$\cdot\left(\frac{1}{2\pi}\int_0^{2\pi}\left|\sum_{k=0}^{n-1}v_k(re^{i\varphi}w_2(re^{i\varphi}))^k\right|^2\,d\varphi\right)^{1/2}.$$

Now, the rest of the proof is an immediate consequence of Littlewood's Theorem 2.1 and Rogosinski's Theorem 2.3.   □

Proof of Theorem 2.9 *of Sheil-Small.* We take $g \prec f \in S$, $g(z) = f(z\omega(z))$, and s as above. Then we may write

$$g(z) = z\omega(z)s(z\omega(z))^2.$$

Taking $w_i = \omega$, $i = 1, 2, 3$, $U = V = s$ in Lemma 2.10 and using the maximum principle we see that the Robertson conjecture is the decisive step needed in the proof of Theorem 2.9.   □

Remark 2.11. It is evident that the Rogosinski conjecture follows from the Sheil-Small inequality (2.9) for $P(z) = z^n$ and that the Bieberbach conjecture is the case $\omega \equiv 1$ of the Rogosinski conjecture.

Remark 2.12. The requirement of Lemma 2.10 about the functions U and V can be considerably relaxed. Namely, it is sufficient to suppose that the functions U and V are holomorphic in a neighbourhood of the origin, only. For such a case the proof is the same except the final step: Instead of Rogosinski's Theorem 2.3 we can apply Theorem 2.4.

2.4   Inverse coefficients

Let Δ be the unit disc $\{z|\,|z| < 1\}$ in the complex plane $\mathbb{C}$ and let S be the usual class of functions holomorphic and univalent in Δ with expansion about the origin,

$$f(z) = z + \sum_{n=2}^{\infty} a_n z^n.$$

We will frequently use the following classical theorem.

Theorem 2.13 (K. Löwner [110]). *If F is the inverse of a function in S and has the expansion*

$$F(w) = w + \sum_{n=2}^{\infty} A_n w^n$$

in a neighbourhood of the origin, then

$$|A_n| \leq \frac{(2n)!}{n!(n+1)!} = \frac{1}{n} \left(\begin{array}{c} 2n \\ n-1 \end{array} \right) \qquad (2.11)$$

with equality only for the inverses of the Koebe functions

$$k_d(z) = \frac{z}{(1+dz)^2}, \quad |d| = 1. \qquad (2.12)$$

By mathematical induction and the Stirling formula one easily gets that

$$\frac{1}{n} \left(\begin{array}{c} 2n \\ n-1 \end{array} \right) = \frac{4^n}{n+1} \frac{(2n-1)!!}{(2n)!!} \leq \frac{4^n}{(n+1)^{3/2}}$$

and that

$$\frac{4^n}{n+1} \frac{(2n-1)!!}{(2n)!!} = \frac{4^n}{(n+1)^{3/2}} \left(\frac{1}{\sqrt{\pi}} + O(1/n) \right) \qquad \text{as } n \to \infty.$$

Here, for $n \in \mathbb{N}$ we use the abbreviations

$$(2n-1)!! = \frac{(2n-1)!}{2^{n-1}((n-1)!)} \quad \text{and} \quad (2n)!! = \frac{(2n)!}{(2n-1)!!}.$$

The classical proof of this Löwner theorem can be found in [78], [70], [128] and [147]. Here we present two generalizations. Each of them implies the theorem. Let K_1 be the inverse of the Koebe function k_1. We have

$$K_1(w) = w + \sum_{n=2}^{\infty} \frac{(2n)!}{n!(n+1)!} w^n,$$

and

$$S(w, K_1) := (K_1''/K_1')' - \frac{1}{2}(K_1''/K_1')^2 = \sum_{n=0}^{\infty} 4^n 6(n+1) w^n,$$

and

$$\log K_1'(w) = \sum_{n=1}^{\infty} b_n w^n,$$

where

$$b_n = \frac{1}{n}\left(\binom{2n}{n} + 2^{2n-1}\right) = \frac{4^n}{n}\left(\frac{1}{2} + \frac{(2n-1)!!}{(2n)!!}\right) \le \frac{4^n}{n}. \tag{2.13}$$

Moreover, one easily gets

$$b_n = \frac{4^{n-1/2}}{n}\left(1 + O(1/\sqrt{n})\right) \quad \text{as } n \to \infty.$$

Since the method to find the expression (2.13) is behind many of the coefficient results of this book, we will give a short proof for (2.13). We use the Cauchy integral formula in the following way. Let

$$\frac{K_1''(w)}{K_1'(w)} = \sum_{n=0}^{\infty}(n+1)b_{n+1}w^n,$$

then

$$(n+1)b_{n+1} = \frac{1}{2\pi i}\int_{k_1(\partial\Delta_r)} \frac{K_1''(w)}{K_1'(w)}\frac{1}{w^{n+1}}\,dw$$

$$= \frac{1}{2\pi i}\int_{\partial\Delta_r}\frac{K_1''(k_1(z))}{K_1'(k_1(z))}\frac{k_1'(z)}{k_1^{n+1}(z)}\,dz = -\frac{1}{2\pi i}\int_{\partial\Delta_r}\frac{k_1''(z)}{k_1'(z)}\frac{1}{k_1^{n+1}(z)}\,dz$$

$$= \frac{1}{2\pi i}\int_{\partial\Delta_r}\frac{(4-2z)(1+z)^{2n+1}}{(1-z)z^{n+1}}\,dz,$$

where $r \in (0,1)$. Hence, we have to find the nth Taylor coefficient of

$$(4-2z)(1+z)^{2n+1}\sum_{k=0}^{\infty}z^k = (1+z)^{2n+1}\left(4 + 2\sum_{k=1}^{\infty}z^k\right).$$

We get from this

$$(n+1)b_{n+1} = 4\binom{2n+1}{n} + 2\sum_{k=0}^{n-1}\binom{2n+1}{k}$$

$$= 2\left(\binom{2n+1}{n} + 2^{2n}\right).$$

This implies

$$b_n = \frac{1}{n}\left(2\binom{2n-1}{n-1} + 2^{2n-1}\right) = \frac{1}{n}\left(\binom{2n}{n} + 2^{2n-1}\right).$$

Consider the following expansions for F, the inverse of a function in S,

$$S(w,F) := (F''/F')' - \frac{1}{2}(F''/F')^2 = \sum_{n=0}^{\infty}C_n w^n$$

and
$$\log F'(w) = \sum_{n=1}^{\infty} B_n w^n.$$

Theorem 2.14 (Klouth and Wirths [91]). *If F is the inverse of a function in S, then $|B_n| \le b_n$ for all $n \ge 1$ and $|C_n| \le 4^n 6(n+1)$ for all $n \ge 0$. Equality for $n \ge 2$ occurs only for the functions $K_d(w) = k_d^{-1}(z)$, $|d| = 1$.*

Since b_n are positive and each A_n is a polynomial with positive coefficients in the B_n, Theorem 2.14 implies Theorem 2.13.

Proof. The proof follows the same line as the classical proof of the Theorem 2.13. We only give here the crucial steps. A function f in S can be embedded into a subordination chain (for details see [128]). It results that the inverse function F has a representation

$$F(w) = \lim_{t \to \infty} \Phi(e^{-t}w, t), \quad \partial \Phi(w,t)/\partial t = w(\partial \Phi(w,t)/\partial w)p(w,t)$$

where $\Phi(w,0) = 0$, $\operatorname{Re} p(w,t) > 0$ and

$$p(w,t) = 1 + \sum_{n=1}^{\infty} p_n(t)w^n$$

for $w \in \Delta$ and $t \ge 0$. Using these and setting

$$L(w,t) := \log \frac{\partial \Phi(w,t)}{\partial w} = \sum_{n=0}^{\infty} B_n(t)w^n,$$

$$M(w,t) := S(w, \Phi(w,t)) = \sum_{n=0}^{\infty} C_n(t)w^n,$$

we get

$$\frac{\partial L}{\partial t} = \frac{\partial L}{\partial w}wp + \frac{\partial(wp)}{\partial w}, \quad \frac{\partial S}{\partial t} = \frac{\partial S}{\partial w}wp + 2S\left(w, \frac{\partial(wp)}{\partial w}\right) + \frac{\partial^3(wp)}{\partial w^3},$$

and, for $n \ge 1$,

$$B_0(t) = t, \quad B_n(t) = \int_0^t e^{n(t-\tau)} \left(\sum_{j=1}^{n-1} jB_j(\tau)p_{n-j}(\tau) + (n+1)p_n(\tau) \right) d\tau,$$

$$C_n(t) = \int_0^t e^{(n-2)(t-\tau)} \left(\sum_{j=1}^{n-1} jC_j(\tau)p_{n-j}(\tau)(2n-j+2) + \frac{(n+3)!}{n!}p_{n+2}(\tau) \right) d\tau.$$

These formulas show that $\operatorname{Re} B_n(t)$, respectively $\operatorname{Re} C_n(t)$, is maximal for a fixed t if and only if we choose $B_n(t), j = 1, \ldots, n-1$, respectively $C_n(t), j = 0, \ldots, n-1$, for $\tau \in [0, t]$ real and maximal and $p_j(\tau) = 2$ in $[0, t]$ for any index j involved in the formulas in question.

Since

$$B_n = \lim_{t \to \infty} e^{-nt} B_n(t), \quad C_n = \lim_{t \to \infty} e^{-(n+2)t} C_n(t),$$

we get that the maximum of $\operatorname{Re} B_n$, respectively $\operatorname{Re} C_n$, is attained if and only if $p(w, t) = (1 + w)/(1 - w)$. Clearly, the assertion of the theorem for $n \geq 1$ follows from the known fact that the problems of finding the maximum of the real part and the maximum of the modulus for the given coefficient are equivalent. To complete the proof we remark that the desired inequality for $C_0 = -6(a_3 - a_2^2)$ is given by the classical inequality $|a_3 - a_2^2| \leq 1$.   □

For a fixed $p \in (0, 1)$, let S_p denote the class of all meromorphic univalent functions f_p in the unit disc Δ with the normalization $f_p(0) = f_p'(0) - 1 = 0$ and $f_p(p) = \infty$. If F_p is the inverse of a function in S_p, then it admits an expansion of the form

$$F_p(w) = w + \sum_{n=2}^{\infty} A_n(p) w^n$$

in a neighbourhood of the origin. In S_p the function

$$\kappa_p(z) = \frac{z}{(1 - pz)\left(1 - \frac{z}{p}\right)}$$

plays the role of the Koebe function in S. Near $w = 0$ the function $K_p(w) = k_p^{-1}(w)$ has the expansion (see, for instance, [33])

$$K_p(w) = w + \sum_{n=2}^{\infty} A_n(p, K_p) w^n$$

with

$$A_n(p, K_p) = \frac{(-1)^{n-1}}{n} \sum_{j=0}^{n-1} \binom{n}{j} \binom{n}{j+1} p^{2j-n+1}.$$

Theorem 2.15 (Baernstein and Schober [33])). *The coefficients of the function F_p satisfy the sharp inequalities*

$$|A_n(p)| \leq A_n(p, K_p).$$

Equality for a single coefficient holds only if $F_p = K_p$.

Clearly, for $p = 1$ Theorem 2.15 gives the Löwner theorem.

The proof of Theorem 2.15 (see [33]) is based on the following integral inequality of Baernstein [32]: *for any $f_p \in S_p$,*

$$\frac{1}{2\pi} \int_0^{2\pi} |f_p(re^{i\theta})|^\alpha d\theta \leq \frac{1}{2\pi} \int_0^{2\pi} |\kappa_p(re^{i\theta})|^\alpha d\theta,$$

whenever $0 < r < 1$ and $-\infty < \alpha < \infty$.

2.5 Domains with bounded boundary rotation

Plane domains with bounded boundary rotation were first studied by Paatero ([122], see also [10]). In Radon's paper [134] one can find general properties and applications of domains whose boundaries are rectifiable curves with bounded rotation.

In the following we need the usual abbreviation: If the functions f and g are analytic in a neighbourhood of the origin, then

$$f(z) = \sum_{k=0}^{\infty} a_k z^k \ll g(z) = \sum_{k=0}^{\infty} b_k z^k$$

if and only if for all $k \in \mathbf{N} \cup \{0\}$ the inequalities

$$|a_k| \le |b_k|$$

are valid.

In [100], Lehto considered the family V_k of functions $f \in S$ such that the boundary rotation of $f(\Delta)$ is at most $k\pi \ge 2\pi$. It is well known that the function

$$f_k(z) = \frac{1}{k}\left[\left(\frac{1+z}{1-z}\right)^{k/2} - 1\right] = z + \sum_{n=2}^{\infty} A_n(k)z^n$$

belongs to V_k.

For any $f \in V_k$ with expansion $f(z) = z + \sum_{n=2}^{\infty} a_n z^n$, it is proved that the inequality

$$|a_n| \le A_n(k)$$

is valid for all $n \ge 2$ (see [45], [2] and [44] and [146]).

The crucial facts are gathered in the following two theorems.

Theorem 2.16 ([45]). *Let φ be a function holomorphic in Δ. If $\alpha \ge 1$ and*

$$\varphi(z) \prec \frac{1+cz}{1-z}$$

for some constant $c \in \overline{\Delta}$, then there exists a probability measure $\mu : [0, 2\pi] \to \mathbb{R}$ such that

$$\varphi(z)^{\alpha} = \int_0^{2\pi} \left(\frac{1 + ce^{it}z}{1 - e^{it}z}\right)^{\alpha} d\mu(t), \quad z \in \Delta.$$

Theorem 2.17. *For any $c \in \overline{\Delta}$ and $\alpha \ge 1$*

$$\left(\frac{1+cz}{1-z}\right)^{\alpha} \ll \left(\frac{1+z}{1-z}\right)^{\alpha}.$$

Consequently, if $\alpha \ge 1$ and

$$\varphi(z) \prec \frac{1+cz}{1-z}$$

for some constant $c \in \overline{\Delta}$, then

$$\varphi(z)^\alpha \ll \left(\frac{1+z}{1-z}\right)^\alpha.$$

Short proofs of these facts are found in [44] and [146]. The original proofs are given in [45] and [2]. In fact, it was shown in [45], that Theorem 2.17 implies the coefficient estimate for the larger class of close-to-convex functions of order $k/2 - 1 \geq 0$ introduced by Ch. Pommerenke [127].

Using Theorem 2.17 and the cited proofs in [44] and [146] we get

Theorem 2.18 ([17]). *If $c \in \overline{\Delta} \setminus \{-1\}$ and $\alpha \geq 1$, then*

$$\frac{1}{c+1}\left(\left(\frac{1+cz}{1-z}\right)^\alpha - 1\right) \ll \frac{1}{2}\left(\left(\frac{1+z}{1-z}\right)^\alpha - 1\right).$$

Proof. For $c \in \overline{\Delta} \setminus \{-1\}$ let $T_c(z) = (1+cz)/(1-z)$. Since $T_c(\Delta)$ is a halfplane whose boundary cuts the real axis in a point of the interval $[0, 1)$ and $1 \in T_c(\Delta)$, there exists, for any $c \in \overline{\Delta} \setminus \{-1\}$ and $\alpha \geq 1$, a $c_1 \in \overline{\Delta}$ such that

$$\varphi_1(z) := T_c(z)^{1-\frac{1}{\alpha}} T_1(z)^{\frac{1}{\alpha}} \prec \frac{1+c_1 z}{1-z}.$$

According to Theorem 2.17 this implies

$$\varphi_1(z)^\alpha \ll T_1(z)^\alpha.$$

Using this and the nonnegativity of the Taylor coefficients of the functions $T_1(z)^\alpha$ and $1/(1-z^2)$ we get

$$\left(\frac{1+cz}{1-z}\right)^{\alpha-1}\frac{1}{(1-z)^2} = \varphi_1(z)^\alpha \frac{1}{1-z^2} \ll \left(\frac{1+z}{1-z}\right)^\alpha \frac{1}{1-z^2}.$$

By integration we obtain the assertion of Theorem 2.18. $\square$

Now, we prove a subordination theorem which we need for the applications. We are concerned with the class of angular domains $\Pi_\alpha = aH_\alpha + b$ ($a \neq 0$) with opening angle $\alpha\pi$, $1 \leq \alpha \leq 2$, which means that there exists a linear transformation $T(z) = az + b$ such that $\Pi_\alpha = T(H_\alpha)$, where

$$H_\alpha = \left\{z \mid |\arg z| < \frac{\alpha\pi}{2}\right\}.$$

Clearly, the assertion of the following theorem is a generalization of the Carathéodory inequality for Taylor coefficients of holomorphic functions with positive real part (see [51], [52], compare also [42], p. 365).

Theorem 2.19 ([17]). *For any angular domain $\Pi_\alpha, 1 \le \alpha \le 2$, and any function g holomorphic in Δ with $g(\Delta) \subset \Pi_\alpha$, the assertion*

$$(g(\zeta) - g(0))\lambda_{\Pi_\alpha}(g(0)) \ll \frac{1}{2\alpha}\left(\left(\frac{1+\zeta}{1-\zeta}\right)^\alpha - 1\right), \quad \zeta \in \Delta,$$

is valid.

Proof. Let

$$G_\alpha(\zeta) := \left(\Phi_{\Pi_\alpha,g(0)}(\zeta) - g(0)\right)\lambda_{\Pi_\alpha}(g(0)), \quad \zeta \in \Delta.$$

The function G_α belongs to the class S and maps Δ univalently onto an angular domain Π_α. This yields the existence of a complex number $c \in \overline{\Delta} \setminus \{-1\}$ such that

$$G_\alpha(\zeta) = \frac{1}{(c+1)\alpha}\left(\left(\frac{1+c\zeta}{1-\zeta}\right)^\alpha - 1\right), \quad \zeta \in \Delta.$$

Therefore, it follows from our earlier assumption that

$$(g(\zeta) - g(0))\lambda_{\Pi_\alpha}(g(0)) \prec G_\alpha(\zeta), \quad \zeta \in \Delta,$$

and in turn that

$$\phi(\zeta) := (1 + (c+1)\alpha(g(\zeta) - g(0))\lambda_{\Pi_\alpha}(g(0)))^{\frac{1}{\alpha}} \prec \frac{1+c\zeta}{1-\zeta}, \quad \zeta \in \Delta.$$

According to Theorem 2.16 this implies that

$$(g(\zeta) - g(0))\lambda_{\Pi_\alpha}(g(0) = \frac{1}{(c+1)\alpha}\int_0^{2\pi}\left(\left(\frac{1+ce^{it}\zeta}{1-e^{it}\zeta}\right)^\alpha - 1\right)d\mu(t), \quad \zeta \in \Delta.$$

This together with Theorem 2.18 yields the result of Theorem 2.19. □

Theorem 2.20 ([17]). *For $1 \le \alpha \le 2$ let*

$$h_\alpha(z) := \frac{1}{2\alpha}\left(1 - \left(\frac{1-z}{1+z}\right)^\alpha\right), \quad z \in \Delta.$$

Let further $h \in S$ and $h(\Delta)$ be an angular domain Π_α. We denote by h^{-1} the function inverse to h and by h_α^{-1} the function inverse to h_α. If we let $h^{-1}(w)$ and $h_\alpha^{-1}(w)$ represent the Taylor expansions of h^{-1} and h_α^{-1} in a neighbourhood of the origin, then

$$h^{-1}(w) \ll h_\alpha^{-1}(w).$$

Proof. Since there exists a complex number $c \in \overline{\Delta} \setminus \{-1\}$ such that any function h of the above type may be written in the form

$$h(z) = \frac{1}{(c+1)\alpha}\left(1 - \left(\frac{1-z}{1+cz}\right)^\alpha\right), \quad z \in \Delta,$$

we get by a straightforward computation that

$$h^{-1}(w) = \frac{1 - (1 - (c+1)\alpha\,w)^{\frac{1}{\alpha}}}{1 + c(1 - (c+1)\alpha\,w)^{\frac{1}{\alpha}}}.$$

Now, we use the expansion

$$(1 - (c+1)\alpha\,w)^{\frac{1}{\alpha}} = \sum_{k=0}^{\infty} (-\alpha)^k \left(\begin{array}{c} \frac{1}{\alpha} \\ k \end{array} \right) (c+1)^k w^k$$

and the fact that, for $k \geq 1$ and $\alpha \geq 1$,

$$D_k(\alpha) := -\frac{(-\alpha)^k}{(c+1)^k} \left(\begin{array}{c} \frac{1}{\alpha} \\ k \end{array} \right) \geq 0.$$

Hence,

$$h^{-1}(w) = \left\{ \sum_{k=1}^{\infty} D_k(\alpha)(c+1)^{k-1} w^k \right\} \left\{ 1 - c \sum_{k=1}^{\infty} D_k(\alpha)(c+1)^{k-1} w^k \right\}^{-1}$$

$$\ll \left\{ \sum_{k=1}^{\infty} D_k(\alpha) 2^{k-1} w^k \right\} \left\{ 1 - \sum_{k=1}^{\infty} D_k(\alpha) 2^{k-1} w^k \right\}^{-1} = h_{\alpha}^{-1}(w).$$

This completes the proof of Theorem 2.20. $\qquad\qquad\qquad\qquad\qquad$ $\square$

Chapter 3

The Poincaré metric

For more than two thousand years, mathematicians and other people believed in
the "truth" of the Euclidean parallel axiom, before Lobachevsky and Bolyai about
1825–1830 found that it is possible to get a new geometry by substituting this
axiom with the following.

*Through a point in the plane not lying on the given straight line one can draw
more than one straight line not intersecting the given line.*

At the beginning, the existence of such a geometry was not accepted by mathematicians, except for C. F. Gauss. In 1868 Beltrami interpreted the Lobachevsky
(or hyperbolic) geometry as the natural geometry on a surface of constant negative
curvature and thus proved that there exists a model for such a geometry. During
this work, he was able to use the theorems of Gauss (1827) and Milding (1840) on
surface geometry. Some years later, in 1871, Klein made this geometry "visible"
to geometers by an interpretation in projective geometry. In accordance with his
model, Klein introduced the term "hyperbolic geometry".

In 1882 Poincaré discovered the following interpretation: the unit disc Δ
equipped with the conformally invariant metric

$$\lambda_\Delta(z)|dz| := \frac{|dz|}{1 - |z|^2}$$

can be regarded as the hyperbolic plane.

3.1 Background

Since any conformal automorphism of the unit disc Δ has the form

$$z = T(\zeta) := e^{i\alpha} \frac{\zeta - a}{1 - \overline{a}\zeta}, \qquad a \in \Delta, \ \alpha \in \mathbb{R},$$

by straightforward computations one gets that the Poincaré metric is conformally invariant in Δ, that is

$$\frac{|dz|}{1 - |z|^2} = \frac{|d\zeta|}{1 - |\zeta|^2} \tag{3.1}$$

for all $z = T(\zeta)$ in Δ.

The hyperbolic distance between two points $z_1, z_2 \in \Delta$ is defined by

$$D_\Delta(z_1, z_2) = \inf \int_\Gamma \lambda_\Delta(z) \, |d\,z|, \tag{3.2}$$

where the infimum is taken over all piecewise smooth curves $\Gamma \subset \Delta$ joining z_1 with z_2. The geodesics (or paths of shortest distance) consist of the images of diameters of Δ under conformal automorphisms T. These are the diameters of Δ and the circular arcs in Δ orthogonal to its boundary $\partial\Delta$. If these arcs are called "straightlines", one has a model of the hyperbolic plane.

Using the "straightlines", one can find that the hyperbolic distance for $z_1, z_2 \in \Delta$ is given in the form

$$D_\Delta(z_1, z_2) = \frac{1}{2} \log \frac{1+\rho}{1-\rho}, \quad \text{where} \quad \rho = \left| \frac{z_1 - z_2}{1 - \overline{z_1} z_2} \right|. \tag{3.3}$$

An alternative interpretation of the hyperbolic geometry is given by the half plane model of Poincaré. In the half plane $H_1 = \{z \in \mathbb{C} | \operatorname{Re} z > 0\}$ the element of hyperbolic arc length is

$$\lambda_{H_1}(z)|dz| = \frac{|dz|}{2x}, \ z = x + iy \in H_1.$$

In this case geodesics are horizontal lines and circles orthogonal to the imaginary axis. Using the conformal map $\zeta = (1 - z)/(1 + z)$ of the unit disc onto the half plane, it is easily seen that these two models are equivalent. The equivalence is given by the simple equation

$$\lambda_{H_1}(\zeta)|d\zeta| = \lambda_\Delta(z)|dz|, \qquad \zeta = \frac{1-z}{1+z}.$$

Now, let Ω denote a domain in $\overline{\mathbb{C}}$ with three or more boundary points in $\overline{\mathbb{C}}$. According to the Riemann mapping theorem, if Ω is simply connected and $z_0 \in \Omega \setminus \{\infty\}$ is fixed, then there exists a unique conformal map f_0 of Δ onto Ω such that

$$f_0(0) = z_0 \quad \text{and} \quad f_0'(0) > 0.$$

This quantity $f_0'(0)$ is called the conformal radius of Ω at the point z_0 and will be denoted in the following by $R(z_0, \Omega)$. If Ω is as above but not simply connected, the generalization of Riemann's mapping theorem due to Poincaré (see, for instance, [3] and [70], p. 255) asserts that there exists a unique universal covering map f_0 of

Δ onto Ω which has the same normalization as the above conformal map. In this case the quantity $f_0'(0)$ is called the hyperbolic radius (see, for instance, [34]).

The Poincaré (or hyperbolic) metric in Ω is defined by the equation

$$\lambda_\Omega(z)|dz| := \frac{|d\zeta|}{1 - |\zeta|^2}, \qquad z = f_0(\zeta) \in \Omega. \tag{3.4}$$

Concerning the universal covering map f_0 for a multiply connected domain, we have to mention that f_0 is a holomorphic or meromorphic function, while its inverse $f_0^{-1}(z)$ is multivalued. But it is well known that any conformal or universal covering map f of Δ onto Ω is given by

$$f(\zeta) = f_0(T(\zeta)) = f_0\left(e^{i\alpha}\frac{\zeta - a}{1 - \overline{a}\zeta}\right), \qquad a \in \Delta, \ \alpha \in \mathbb{R}.$$

This together with (3.1) imply that the density $\lambda_\Omega(z)$ is well-defined by equation (3.4), i.e., it depends neither on the choice of the mapping function nor on the choice of the branch of f^{-1}. Further, the metric $\lambda_\Omega(z)|dz|$ is conformally invariant.

From the above definitions it follows that

$$R(z, \Omega) = \frac{1}{\lambda_\Omega(z)} = |f'(\zeta)|(1 - |\zeta|^2), \tag{3.5}$$

where f is a conformal or universal covering map of Δ onto Ω and $z = f(\zeta)$, $\zeta \in \Delta$, $z \in \Omega$.

Taking partial derivatives of R by using (3.5) and the Wirtinger calculus

$$2\frac{\partial R(z, \Omega)}{\partial z} = \frac{\partial R(z, \Omega)}{\partial x} - i\frac{\partial R(z, \Omega)}{\partial y}, \ z = x + iy \in \Omega,$$

$$2\frac{\partial R(z, \Omega)}{\partial \overline{z}} = \frac{\partial R(z, \Omega)}{\partial x} + i\frac{\partial R(z, \Omega)}{\partial y} =: \nabla R(z, \Omega), \ z = x + iy \in \Omega,$$

leads to the formulas

$$\frac{\partial R(z, \Omega)}{\partial z} = \frac{|f'(\zeta)|}{f'(\zeta)}\left(\frac{1 - |\zeta|^2}{2}\frac{f''(\zeta)}{f'(\zeta)} - \overline{\zeta}\right) \tag{3.6}$$

and

$$R(z, \Omega)\frac{\partial^2 R(z, \Omega)}{\partial \overline{z}\partial z} = \left|\frac{\partial R(z, \Omega)}{\partial \overline{z}}\right|^2 - 1.$$

Hence, the conformal (or hyperbolic) radius $R = R(z, \Omega)$ satisfies Liouville's equation

$$R\triangle R = |\nabla R|^2 - 4, \tag{3.7}$$

where $R = R(z, \Omega)$, $\triangle R$ denotes the Laplacian of R and ∇R its gradient.

The most known forms of the equation (3.7) are the following nonlinear elliptic equation

$$\triangle u = -4e^{-2u},$$

where $u = u(x, y) := -\log \lambda_\Omega(z) = \log R(z, \Omega)$ for $z = x + iy \in \Omega$, and an equivalent formula

$$K = e^{2u} \triangle u = -4$$

that defines the Gaussian curvature of the metric. Thus, the Liouville equation for R is equivalent to the fact that the hyperbolic metric defined as above has the Gaussian curvature $K = -4$ (see, for instance, [3], [34] and [70]).

Remark 3.1. Clearly, to deal with the hyperbolic metric with Gaussian curvature $K = -1$, one has to choose the density of the metric as $2\lambda_\Omega(z)$, where $\lambda_\Omega(z)$ is defined as above.

3.2   The Schwarz-Pick inequality

We begin by the classic Schwarz lemma.

Theorem 3.2 (Schwarz's lemma). *Let f be holomorphic in the unit disc Δ. If $f(0) = 0$ and $|f(z)|$ is bounded by 1 in Δ, then*

$$|f(z)| \leq |z|, \quad z \in \Delta, \tag{3.8}$$

and

$$|f'(0)| \leq 1 \tag{3.9}$$

with equality in (3.8) for some $z \neq 0$ or in (3.9) occurring if and only if $f(z) = cz$ for some unimodular complex constant c.

As is indicated in [3], p. 21, the Schwarz lemma and its classic proof are due to Carathéodory [53]; Schwarz proved it only for univalent mappings [149]. In [124], [125] Pick realized the invariant character of Schwarz's lemma.

Theorem 3.3 (Pick [125], see also Lindelöf [105]). *A holomorphic mapping f of the unit disc Δ into itself satisfies the inequalities*

$$\left| \frac{f(z_1) - f(z_2)}{1 - \overline{f(z_1)} f(z_2)} \right| \leq \left| \frac{z_1 - z_2}{1 - \overline{z_1} z_2} \right|, \quad z_1, z_2 \in \Delta, \tag{3.10}$$

and

$$\frac{|f'(z)|}{1 - |f(z)|^2} \leq \frac{1}{1 - |z|^2}, \quad z \in \Delta. \tag{3.11}$$

Nontrivial equality holds if and only if f is a conformal automorphism of Δ, i.e., $f(z) = e^{i\alpha}(z - a)/(1 - \overline{a}z), \qquad a \in \Delta, \ \alpha \in \mathbb{R}$.

It is evident that inequalities (3.10) and (3.11) correspond to Schwarz's inequalities (3.8) and (3.9), respectively. Geometrically, inequality (3.10) means that any holomorphic mapping f of the unit disc Δ into itself decreases the hyperbolic distance between two points. Also, (3.11) implies that f decreases the hyperbolic lengths of an arc and the hyperbolic area of a set.

The following assertion is called the hyperbolic metric principle (see, for instance, [70] and [120]).

Theorem 3.4. *Let Ω and Π be domains in the extended complex plane such that each of them has at least three boundary points. If $f \in A(\Omega, \Pi)$, then*

(i) *$L(f(\gamma)) \le L(\gamma)$ for any rectifiable curve $\gamma \subset \Omega$ and its image $f(\gamma) \subset \Pi$ with length $L(\gamma)$ and $L(f(\gamma))$ in the hyperbolic metric of Ω and Π, respectively; equality holds if and only if $f(z) = f_2(f_1^{-1}(z))$, where f_1 and f_2 are conformal (universal covering) maps of the unit disc onto Ω and Π, respectively.*

(ii) *the differential length elements at $z \in \Omega$ and at its image $w = f(z) \in \Pi$ satisfy the inequality*

$$\lambda_\Pi(w)|dw| \le \lambda_\Omega(z)|dz| \qquad (3.12)$$

with the same condition for equality.

We shall call the Schwarz-Pick inequality the equation (3.12) written in the form

$$|f'(z)| \le \frac{\lambda_\Omega(z)}{\lambda_\Pi(f(z))}. \qquad (3.13)$$

The Schwarz-Pick inequality applied to the function $f \in A(\Omega', \Omega)$ defined by $\zeta = f(z) = z$, $z \in \Omega'$, immediately gives the following comparison of densities.

Theorem 3.5. *If $\Omega' \subset \Omega$, then $\lambda_{\Omega'}(z) \ge \lambda_\Omega(z)$ for any point $z \in \Omega'$.*

Explicit formulas for the Poincaré density are known for several domains. We shall present three of them. They will be useful in applications of the comparison theorem 3.5.

1. For angular domains $H_\alpha = \{z \in \mathbb{C} \setminus \{0\} \mid |\arg z| < \frac{\alpha\pi}{2}\}$, $\alpha \in (0, 2]$, one easily gets

$$1/\lambda_{H_\alpha}(z) = R(z, H_\alpha) = 2\,\alpha r \cos\frac{\theta}{\alpha}, \quad z = re^{i\theta} \in H_\alpha,$$

and

$$1/\lambda_{H_0}(z) = R(z, H_0) = \frac{4}{\pi}\cos\frac{\pi y}{2}, \quad z = x + iy, \ |y| < 1$$

for the strip $H_0 = \{z \in \mathbb{C} \mid |\text{Im } z| < 1\}$.

2. For the punctured unit disc $\Delta' = \{z \in \mathbb{C} \mid 0 < |z| < 1\}$,

$$1/\lambda_{\Delta'}(z) = R(z, \Delta') = 2|z| \log\frac{1}{|z|}, \quad z \in \Delta'.$$

3. It is a much harder problem to find an explicit formula for the density of the domain $\mathbb{C} \setminus \{0,1\} = \overline{\mathbb{C}} \setminus \{0,1,\infty\}$. The following formula is due to Agard [1]:

$$\frac{1}{\lambda_{\mathbb{C} \setminus \{0,1\}}(z)} = \frac{|z||z-1|}{\pi} \int\!\!\int_{\mathbb{C}} \frac{du\,dv}{|w||w-1||w-z|}, \quad w = u + iv.$$

The classical representation formula for $\lambda_{\mathbb{C} \setminus \{0,1\}}$ is given by

$$\lambda_{\mathbb{C} \setminus \{0,1\}}(z) = \frac{1}{2|\lambda'(\tau)|\,\mathrm{Im}\,\tau} = \frac{|g'(z)|}{2\,\mathrm{Im}\,g(z)},$$

where $\lambda(\tau)$ is the elliptic modular function, g is its inverse, $\tau \in \Lambda := \{\tau \mid \mathrm{Im}\,\tau > 0\}$ and $z \in \mathbb{C} \setminus \{0,1\}$. The modular function $\lambda : \Lambda \to \mathbb{C} \setminus \{0,1\}$ is a universal covering map which maps the triangle $\{\tau \in \Lambda \mid 0 < \mathrm{Re}\,\tau < 1, |\tau - 1/2| > 1/2\}$ onto Λ in such a way that $\lambda(0) = 1$, $\lambda(1) = \infty$, $\lambda(\infty) = 0$, and

$$\lambda(\tau) = 16\,q \prod_{n=1}^{\infty} \left(\frac{1 + q^{2n}}{1 + q^{2n-1}} \right)^{8}, \quad q = e^{\pi i \tau} \tag{3.14}$$

(see, for instance, [107]).

The Liouville equation for the hyperbolic density $\lambda_{\mathbb{C} \setminus \{0,1\}}(z)$ and formula (3.14) are used in a proof of some monotonicity properties of the metric.

Theorem 3.6 (see A. Bermant [39], J. A. Hempel [80] and compare also A. Yu. Solynin and M. Vuorinen [154] for further results). *Let $z = x + iy = re^{i\theta}$. The Poincaré density $\lambda_{\mathbb{C} \setminus \{0,1\}}(z)$ has the following properties:*

$$y \frac{\partial \lambda_{\mathbb{C} \setminus \{0,1\}}(z)}{\partial y} < 0, \quad y \neq 0, \tag{3.15}$$

$$\theta \frac{\partial \lambda_{\mathbb{C} \setminus \{0,1\}}(z)}{\partial \theta} < 0, \quad 0 < |\theta| < \pi, \tag{3.16}$$

$$\left(\frac{\partial \lambda_{\mathbb{C} \setminus \{0,1\}}(z)}{\partial r} + \frac{\lambda_{\mathbb{C} \setminus \{0,1\}}(z)}{r} \right) \log r < 0, \quad r \neq 1. \tag{3.17}$$

There are many generalizations and applications of the classical Theorems 3.2 and 3.3. Especially, there are close relations between Theorems 3.2 and 3.3 in the theory of unimodular bounded holomorphic functions due to Carathéodory, Denjoy, Julia and Wolff. We cite the following statements, which sometimes are quoted as **the Grand Iteration Theorem** (see [150], p. 78 and compare also [131], p. 82, and [55]).

Theorem 3.7. *Let φ be a holomorphic self-map of Δ that is not a conformal automorphism of Δ with a fixed point in $\overline{\Delta}$. Then there is a unique point $\omega \in \overline{\Delta}$ such that the iterates*

$$\varphi^{n}(z) = (\varphi \circ \cdots \circ \varphi)(z)$$

converge to ω as $n \to \infty$ uniformly on any compact subset of Δ and

(C$_1$) *if* $\omega \in \Delta$, *then* φ *has no fixed point in* $\Delta \setminus \{\omega\}$, *and*

$$\varphi(\omega) = \omega, \, 0 \leq |\varphi'(\omega)| < 1,$$

(C$_2$) *if* $\omega \in \partial\Delta$ *(Denjoy-Wolff point of* φ*), then* φ *has no fixed point in* Δ *and* φ *and* φ' *have angular limits at* ω *such that*

$$\varphi(\omega) = \omega, \, 0 < \varphi'(\omega) \leq 1,$$

where

$$\varphi'(\omega) = \sup_{z \in \Delta} \left\{ \operatorname{Re} \frac{\omega + z}{\omega - z} \right\} \left\{ \operatorname{Re} \frac{\omega + \varphi(z)}{\omega - \varphi(z)} \right\}^{-1}.$$

For further deep results on the boundary behavior of bounded holomorphic functions we refer the reader to the books by L. V. Ahlfors [3], by J. B. Garnett [66] and by Ch. Pommerenke [131] (see also R. B. Burckel [48] and G. M. Goluzin [70] for other developments and applications of Theorems 3.2 and 3.3.)

3.3 Estimates using the Euclidean distance

Choosing Ω' as the disc with center z and radius $\delta(z) = \operatorname{dist}(z, \partial\Omega)$ in Theorem 3.5, one gets

Theorem 3.8. *If* $\delta(z)$ *denotes the distance from* $z \in \Omega$ *to the boundary* $\partial\Omega$ *of* Ω, *then* $\lambda_\Omega(z) \leq 1/\delta(z)$ *for any* $z \in \Omega$.

The next assertion is known as the Koebe 1/4-theorem (for a proof see [60], [70], [78] or [128]).

Theorem 3.9. *If* Ω *is a simply connected domain in* $\mathbb{C}$, *then* $\delta(z)\lambda_\Omega(z) \geq \frac{1}{4}$ *for any* $z \in \Omega$.

We will need Landau's theorem on holomorphic functions that omit two fixed values. More precisely, the Landau theorem concerns functions

$$f(z) = \sum_{n=0}^{\infty} a_n z^n$$

holomorphic in the unit disc Δ and omitting the values 0 and 1 in Δ. By our notation,

$$f \in A(\Delta, \mathbb{C} \setminus \{0, 1\}).$$

In the sequel, the known constant

$$\frac{1}{2\lambda_{\mathbb{C}\setminus\{0,1\}}(-1)} = i\lambda'(1+i) = \frac{\Gamma(1/4)^4}{4\pi^2} = 4.3768796\ldots$$

is used. J. A. Hempel and J. A. Jenkins established an explicit sharp bound in the Landau theorem.

Theorem 3.10 (Landau; see J. A. Hempel [80] and J. A. Jenkins [87] for proofs).
If the function f is holomorphic and omits 0 and 1 in Δ, then

$$|a_1| \leq 2|a_0| \left\{ |\log |a_0|| + \frac{\Gamma(1/4)^4}{4\pi^2} \right\}. \tag{3.18}$$

Equality holds only for the universal covering map $f : \Delta \to \mathbb{C}\backslash\{0,1\}$ with $a_0 = -1$.

Proof by J. A. Hempel [80]. By the hyperbolic metric principle we have the sharp inequality

$$|a_1| = |f'(0)| \leq 2/\rho(a_0), \quad \rho(z) := 2\lambda_{\mathbb{C}\backslash\{0,1\}}(z).$$

Thus, Theorem 3.10 states that

$$\frac{1}{\lambda_{\mathbb{C}\backslash\{0,1\}}(z)} = \frac{2}{\rho(z)} \leq 2|z| \left\{ |\log |z|| + \frac{1}{\rho(-1)} \right\}. \tag{3.19}$$

According to formula (3.15) of Theorem 3.6, to prove inequality (3.19) it is sufficient to consider real values of $z = x + iy = re^{i\theta}$ lying in the interval $(-\infty, 0)$. Let us introduce the real function

$$w(\sigma) := u(e^{\sigma + i\pi}) + \sigma, \quad \sigma := \log r, \, u := \log \rho.$$

Equation (3.19) is equivalent to

$$e^{-w(\sigma)} \leq |\sigma| + e^{-w(0)}. \tag{3.20}$$

From (3.17) of Theorem 3.6 it follows that $\sigma w'(\sigma) < 0$ for $\sigma \neq 0$. Again from (3.16) of Theorem 3.6 we deduce that $\partial^2 u / \partial\theta^2 \geq 0$ for $\theta = \pi$. This together with Liouville's equation

$$\frac{\partial^2 u}{\partial\sigma^2} + \frac{\partial^2 u}{\partial\theta^2} = e^{2(\sigma+u)}$$

give

$$\partial^2 u / \partial\sigma^2 \leq e^{2(\sigma+u)}, \quad \theta = \pi,$$

which is equivalent to the inequality

$$w'' \leq e^{2w}.$$

Integrating this inequality in $(-\infty, 0]$ and $[0, \infty)$ using the local behavior $w' \to 0$ and $e^{2w} = r^2\rho^2(z) \to 0$ as $\sigma \to \infty$ (see [80] for details), we have (3.20).
    This completes the proof of Theorem 3.10. $\square$

Remark 3.11. Since the function

$$1 - f(z) = 1 - a_0 - a_1 z - \sum_{n=2}^{\infty} a_n z^n$$

also omits 0 and 1, we can take $1 - a_0$ in (3.18) instead of a_0. Consequently, Theorem 3.10 implies

$$|a_1| \leq \min_{\zeta \in \{a_0, 1 - a_0\}} 2|\zeta| \left\{ |\log |\zeta|| + \frac{\Gamma(1/4)^4}{4\pi^2} \right\}.$$

We shall consider an important application of Landau's theorem in the theory of uniformly perfect sets.

Let $\Omega \subset \mathbb{C}$ be an open set with more than one boundary point in $\mathbb{C}$. The annulus

$$An(r, s; a) = \{ z \mid r < |z - a| < s \} \subset \Omega$$

with center a is said to separate the compact set

$$E = \overline{\mathbb{C}} \setminus \Omega$$

in the Riemann sphere, whenever both components of $\overline{\mathbb{C}} \setminus An(r, s; a)$ have non-empty intersections with E. Since in the next section we are concerned with conformal images of such annuli, called conformal annuli, we will characterise annuli of the above form as genuine annuli.

Following Ch. Pommerenke (see [129] and also [36], [37], [55], [65], [130]), we define the maximum modulus $M_0(\Omega)$ as the supremum of the moduli of all annuli $An \subset \Omega$ such that An separates E and the center a of An belongs to $\partial\Omega$. As usual the modulus of an annulus An is defined by

$$M(An(r, s; a)) = \frac{1}{2\pi} \log \frac{s}{r}.$$

We take $M_0(\Omega) = 0$, if Ω contains no genuine annulus centered at a boundary point and separating E.

Theorem 3.12 (Beardon and Pommerenke, compare [37] and [129]). *Let $\Omega \subset \mathbb{C}$ be an open set with more than one boundary point in $\mathbb{C}$. If $\lambda_\Omega(z)$ is the density of the Poincaré metric defined in each component of Ω with curvature $K = -4$ and $M_0(\Omega)$ is finite, then for any $z \in \Omega$,*

$$\frac{1}{2\lambda_\Omega(z) \operatorname{dist}(z, \partial\Omega)} \leq \pi M_0(\Omega) + \frac{(\Gamma(1/4))^4}{4\pi^2}. \tag{3.21}$$

Proof. Consider the function

$$\beta_\Omega(z) := \min \left\{ \left| \log \left| \frac{z - a}{b - a} \right| \right| \, \Big| \, |z - a| = \operatorname{dist}(z, \partial\Omega), a \in \partial\Omega, \ b \in \partial\Omega \right\}.$$

We fix z in Ω and choose points a and b in $\partial\Omega$ such that

$$|z - a| = \operatorname{dist}(z, \partial\Omega), \quad \beta_\Omega(z) = \left| \log \left| \frac{z - a}{b - a} \right| \right|.$$

Applying the comparison theorem 3.5 and using the function $g(w) = (w-a)/(b-a)$, $w \in \Omega$, one obtains

$$\lambda_{\mathbb{C}\setminus\{0,1\}}(g(z)) \leq \lambda_{g(\Omega)}(g(z)) = |b-a|\lambda_\Omega(z)$$

which is equivalent to

$$\frac{1}{2\lambda_\Omega(z)\,\mathrm{dist}(z,\partial\Omega)} \leq \frac{1}{2|g(z)|\lambda_{\mathbb{C}\setminus\{0,1\}}(g(z))}.$$

This together with Landau's inequality (3.19) give

$$\frac{1}{2\lambda_\Omega(z)\,\mathrm{dist}(z,\partial\Omega)} \leq \beta_\Omega(z) + \frac{1}{\rho(-1)}. \tag{3.22}$$

Consider now any point $a' \in \Omega$ such that $|z-a'| = \delta$, where $\delta = \mathrm{dist}(z,\partial\Omega)$. Clearly, if the circle $\{w \mid |w-a'| = \delta\}$ meets $\partial\Omega$, then $\beta_\Omega(z) = 0$. Otherwise there exists a maximal non-empty annulus of the form

$$An = \{w \mid \delta e^{-m} < |w-a'| < \delta e^m\},$$

such that

$$a' \in \partial\Omega, \quad An \subset \Omega, \quad m/\pi \leq M_0(\Omega).$$

As An is maximal, there is a point $b' \in \partial\Omega \cap \partial An$, consequently

$$\beta_\Omega(z) \leq \left|\log\left|\frac{z-a'}{b'-a'}\right|\right| = m \leq \pi M_0(\Omega).$$

This together with inequality (3.22) give inequality (3.21).

The proof of Theorem 3.12 is complete. $\square$

Theorems 3.10 and 3.12 are connected with applications of the Poincaré metric in problems of function theory and of mathematical physics. As an example we mention a result on Hardy type inequalities.

Let $C_0^\infty(\Omega)$ be the usual family of smooth functions with compact support in Ω. For $p \in [1,\infty)$ we shall consider the Hardy constant $c_p(\Omega)$ defined by

$$c_p(\Omega) = \sup\left\{\left\|f/\delta^{2/p}\right\|_{L^p(\Omega)} \mid f \in C_0^\infty(\Omega), \left\|(\nabla f)/\delta^{2/p-1}\right\|_{L^p(\Omega)} = 1\right\},$$

where $\delta = \mathrm{dist}(x+iy,\partial\Omega)$.

Theorem 3.13 (Avkhadiev [12], see also [13]). *Let $\Omega \subset \mathbb{C}$ be an open set with more than one boundary point in $\mathbb{C}$. For any $p \in [1,\infty)$,*

$$\min\{2,p\}M_0(\Omega) \leq c_p(\Omega) \leq 2p\left(\pi M_0(\Omega) + \frac{(\Gamma(1/4))^4}{4\pi^2}\right)^2. \tag{3.23}$$

The proof of the upper bound in (3.23) uses the inequality (3.21). The key role in this proof is played by the following lemma.

Lemma 3.14 (see [12], p. 10 and [13], Theorem 3). *Let $\Omega \subset \overline{\mathbb{C}}$ be an open set of the Riemann sphere with more than two boundary points in $\overline{\mathbb{C}}$. If $1 \le p < \infty$, then for any $f \in C_0^\infty(\Omega)$,*

$$\int\int_\Omega |f|^p \lambda_\Omega^2 \, dx \, dy \le \left(\frac{p}{2}\right)^p \int\int_\Omega \frac{|\nabla f|^p}{\delta^{2-p}} (\lambda_\Omega \delta)^{2-2p} \, dx \, dy, \qquad (3.24)$$

where $\delta = \operatorname{dist}(x + iy, \partial\Omega)$ and $\lambda_\Omega = \lambda_\Omega(x + iy)$.

Remark 3.15. In [13], the formula (4) contains a misprint. Namely, the power $2p - 2$ has to be replaced by $2 - 2p$.

The inequality (3.24) is a generalization of a known fact for simply or doubly connected domains. Namely, if any component of Ω can be mapped conformally onto either the unit disc Δ or an annulus of the form $\{z \mid q < |z| < 1\}$, then for any $f \in C_0^\infty(\Omega)$,

$$\int\int_\Omega |f|^p \lambda_\Omega^2 \, dx \, dy \le \left(\frac{p}{2}\right)^p \int\int_\Omega |\nabla f|^p \lambda_\Omega^{2-p} \, dx \, dy. \qquad (3.25)$$

The inequality (3.25) can be derived using the original Hardy inequality and the conformal invariance of the hyperbolic metric (see, for instance, [4], [11], [12], [13], [63]). Unfortunately, in the general case, where no constraints on the components of Ω are imposed, inequality (3.25) does not hold even for $(p/2)^p$ replaced by any other finite constant. For instance, an inequality of the form (3.25) does not hold for $\Omega = \mathbb{C} \setminus \{0, 1\}$.

3.4 An application of Teichmüller's theorem

As is observed in the book [55] of Carleson and Gamelin (see p. 64):

A simple scaling and normal families argument shows that conformal annuli of large modulus contain genuine annuli of large modulus.

As usual, the modulus $M(D)$ of a conformal annulus, or a doubly connected domain, equals the modulus of the genuine annulus that is conformally equivalent to D (see for instance [3] and [97]).

Let $\Omega \subset \mathbb{C}$ be an open set with more than one boundary point in $\mathbb{C}$. A compact set $E = \overline{\mathbb{C}} \setminus \Omega$ is said to be uniformly perfect if

$$M(\Omega) = \sup\{M(D) \mid D \subset \Omega \text{ doubly connected and separating } E\}$$

is finite. The most known quantity describing the geometry of uniformly perfect sets was defined by Pommerenke in [129]: *A compact set E on the Riemann sphere*

$\overline{\mathbb{C}}$ *that contains the point at infinity is uniformly perfect if there exists a constant* $c \in (0, 1)$ *such that, for every* $z_0 \in E \setminus \{\infty\}$ *and every* $r \in (0, \infty)$, *the set* $E \cap$ $An(cr, r; z_0)$ *is not empty.*

Let $C_0(E)$ be the supremum of the admissible constants $c \in (0, 1)$ for a uniformly perfect set. It is evident that

$$C_0(E) = e^{-2\pi M_0(\Omega)}, \quad \Omega = \mathbb{C} \setminus E,$$

where $M_0(\Omega)$ is the quantity defined in Section 3.3 as the supremum of the moduli of all genuine annuli $An \subset \Omega$ such that An separates E and the center of An belongs to $\partial\Omega$ (see [129] and also [36], [37], [55], [65], [130]).

We take $M(\Omega) = 0$ for open sets that have simply connected components only and $M_0(\Omega) = 0$, if Ω contains no genuine annulus centered at a point of $\partial\Omega$ and separating E.

Since any genuine annulus is a conformal annulus, the inequality

$$M_0(\Omega) \leq M(\Omega) \tag{3.26}$$

is trivial and the above remark of Carleson and Gamelin implies

$$M(\Omega) = \infty \quad \Longrightarrow \quad M_0(\Omega) = \infty. \tag{3.27}$$

On the other hand, using his theorem on extremal moduli, Teichmüller proved in [162] that any doubly connected domain D with $M(D) > 1/2$ contains a circle which separates the components of its complement. Moreover, the center of this circle may be chosen at E_1, the bounded component of $\mathbb{C} \setminus D$, and the radius of the circle equals the diameter of E_1.

Solynin (see [153]) and Herron, Liu, and Minda (see [82]) proved that a separating circle exists for doubly connected domains D with $M(D) > 1/4$, if one does not restrict the position of the center of the circle. In [82] it is also proved that such D contain separating annuli An of modulus $M(An) = M(D) - c$ ($c \approx 0.46$), where again no restriction is imposed on the center of the annulus. In the same paper, they prove that $M(An) \geq M(D) - \tilde{c}$ ($\tilde{c} = \frac{1}{\pi} \log(2\sqrt{2} + 2) \approx 0.501$) if one fixes the center of An on ∂D.

In the following we will use the result of Teichmüller to derive a new quantitative version of the Carleson-Gamelin remark, where we find $1/2$ to be the best possible constant in a similar inequality, not only choosing the center of An on ∂D but fixing it as Teichmüller did.

We will apply this result to different characterizations of uniformly perfect sets and to estimates of the second derivative of functions analytic on open sets with uniformly perfect boundaries. In addition to (3.26) and (3.27) we prove that

$$M(\Omega) \leq M_0(\Omega) + \frac{1}{2} \tag{3.28}$$

for any open set $\Omega \subset \mathbb{C}$. The proof is based on the above cited theorem of Teichmüller and on a related formula of Ahlfors in [3].

We are concerned with the following theorem of Teichmüller (see [162] and also [3]).

Theorem 3.16. *Of all doubly connected domains that separate the pair $\{-1, 0\}$ from a pair $\{w_0, \infty\}$ with $|w_0| = R$, the one with the biggest modulus is the complement of the union of the segments $[-1, 0]$ and $[R, \infty]$.*

Clearly, to prove the inequality (3.28) for any open set Ω it is sufficient to consider the special case where Ω is a doubly connected domain. Hence, the inequality (3.28) is a corollary of the following extension of Teichmüller's result on the existence of a separating circle.

Theorem 3.17 ([27]). *Any doubly connected domain $D \subset \mathbb{C}$ with $M(D) > 1/2$ contains an annulus $An = An(r, s; z_0)$ such that $M(An) = M(D) - 1/2$, the center z_0 belongs to the bounded component E_1 of $\overline{\mathbb{C}} \setminus D$ and $\operatorname{diam} E_1 = r$. The constant $1/2$ is sharp.*

Proof. Let $E = \overline{\mathbb{C}} \setminus D = E_1 \cup E_2$, where E_1 and E_2 are the connected components of E. Without loss of generality we may assume that the diameter of E_1 equals 1 and that the points $z = 0$ and $z = -1$ belong to E_1. Now, we consider the annulus $An(1, m; 0)$, where $m = \exp(2\pi(M(D) - 1/2))$. It has the following properties: $An(1, m; 0) \cap E_1 = \emptyset$, the center of $An(1, m; 0)$ belongs to E_1, and $M(An(1, m; 0)) = M(D) - 1/2$.

If $An(1, m; 0) \subset D$, then $An(1, m; 0)$ separates E and there is nothing to prove. Let us assume that this is not true, i.e., that there exists a point $w_0 \in E_2$ such that $1 < |w_0| < m$. Taking $|w_0| = R$ we consider the Teichmüller annulus

$$D(R) = \mathbb{C} \setminus ([-1, 0] \cap [R, \infty))$$

with the modulus $\Lambda(R)$ (compare [3]). According to Theorem 3.16 the inequality

$$M(D) \leq \Lambda(R) \tag{3.29}$$

is valid. On the other hand, we will prove that the assumed condition $1 < R < m$ implies that

$$\Lambda(R) < \frac{1}{2\pi} \log R + \frac{1}{2} < M(D). \tag{3.30}$$

Clearly, (3.30) contradicts (3.29). Hence, to complete the proof of Theorem 3.17 it is sufficient to show that the first inequality in (3.30) holds for any $R > 1$. The function $\Lambda(R)$ is implicitly defined by the following formula due to Ahlfors (see [3], p. 75, also compare formula (3.14)):

$$R = \frac{1}{16 q} \prod_{n=1}^{\infty} \left(\frac{1 - q^{2n-1}}{1 + q^{2n}} \right)^8, \quad q = e^{-2\pi\Lambda(R)}. \tag{3.31}$$

It is known that $\Lambda(1) = 1/2$ and that $\Lambda(R) \to \infty$ as $R \to \infty$. If one considers the first equation in (3.31) as the definition of a function $R(q)$, it is evident that $qR(q)$

is a decreasing function of q for

$$0 < q < e^{-\pi} = e^{-2\pi\Lambda(1)}.$$

The function $qR(q)$ decreases from $1/16$ to $\exp(-\pi)$, when q increases from 0 to $\exp(-\pi)$. In particular,

$$16 < \frac{e^{2\pi\Lambda(R)}}{R} < e^{\pi} = 23.14\ldots.$$

for any $R > 1$, and (3.30) follows.

The sharpness of the constant $1/2$ is a consequence of Teichmüller's considerations, since in the case $M(D) = 1/2$ there does not exist a separating annulus with the above properties.

This completes the proof of Theorem 3.17. $\square$

In contrast to the conformal characteristic $M(\Omega)$ the quantity $M_0(\Omega)$ is not a conformal invariant. Nevertheless, it may be useful to remark the following consequence of Theorem 3.17.

Corollary 3.18. *Let Ω_1 and Ω_2 be two conformally equivalent domains in $\mathbb{C}$. If $M_0(\Omega_1)$ is finite, then*

$$|M_0(\Omega_1) - M_0(\Omega_2)| \leq \frac{1}{2}.$$

Remark 3.19. For any genuine annulus An, the inequality

$$M_0(An) < M(An)$$

holds, since we consider only those separating genuine annuli in An, which have a center lying on ∂An. Geometrically it is evident that a genuine annulus An has such a separating annulus, if and only if

$$\frac{1}{2\pi}\log 3 < M(An) \leq M_0(An) + \frac{1}{2\pi}\log 3.$$

3.5   Domains with uniformly perfect boundary

There are about twenty characterizations of domains with uniformly perfect boundary via the Hayman-Wu condition, domains with strong barrier, local behavior of harmonic measure or logarithmic capacity, etc. (see, for instance, [9], [36], [37], [55], [62], [65], [67], [76], [77], [79], [129], [130]).

In this section we continue to use the Pommerenke characteristic $C_0(E) = \exp(-2\pi M_0(\Omega))$, $\Omega = \mathbb{C} \setminus E$. Our aim is to consider the problem of comparing the Euclidian geometry characteristic $M_0(\Omega)$ of open sets with uniformly perfect boundary with the following characteristics of the hyperbolic geometry on Ω:

$$\alpha(\Omega) = \inf\{\lambda_\Omega(z)\mathrm{dist}(z, \partial\Omega) \mid z \in \Omega\} \qquad (3.32)$$

and

$$\gamma(\Omega) = \sup \left\{ |\nabla \lambda_\Omega^{-1}(z)| \mid z \in \Omega \right\}, \tag{3.33}$$

where λ_Ω is the density of the Poincaré metric with curvature -4 defined on the components of Ω.

We will prove that

$$\sup\{M_0(\Omega)\alpha(\Omega)\} = \frac{1}{4} \tag{3.34}$$

and that

$$\sup \left\{ \frac{M_0(\Omega)}{\gamma(\Omega)} \right\} = \frac{1}{4}, \tag{3.35}$$

where the supremum is taken with respect to all open sets $\Omega \subset \mathbb{C}$ that have more than one boundary point in $\mathbb{C}$. In the proof of (3.34) and (3.35) we show that

$$\frac{(\gamma(An))^2}{16} = \frac{1}{4} + (M(An))^2$$

for any genuine annulus An and that

$$4M(An) \sim \frac{1}{\alpha(An)}$$

for genuine annuli if $M(An) \to \infty$.

We must confess that the above considerations have their origin in the central question of this book, compare section 1.2. Let Ω and Π be open sets in $\mathbb{C}$ or in $\overline{\mathbb{C}}$ and let $A(\Omega, \Pi)$ be the set of functions $f : \Omega \to \Pi$ locally holomorphic or meromorphic and in general multivalued. What can be said about the influence of the geometric properties of Ω and Π on the quantities

$$C_n(\Omega, \Pi) = \sup \left\{ \frac{|f^{(n)}(z)|}{n!} \frac{\lambda_\Pi(f(z))}{(\lambda_\Omega(z))^n} \mid z \in \Omega, \ f \in A(\Omega, \Pi) \right\}?$$

The above considerations give us the possibility to determine bounds for $C_2(\Omega, \Pi)$ in terms of the quantities $M_0(\Omega)$ and $M_0(\Pi)$ that are "visible" in Euclidean geometry. During these proofs (see Chapter 6) we get a sharp form of the Osgood-Jørgensen inequality (see [88] and [121]), namely

$$\sup\{\sup\{|\nabla \log \lambda_\Omega(z)| \mathrm{dist}(z, \partial\Omega) \mid z \in \Omega\}\} = 2, \tag{3.36}$$

where the first supremum is taken with respect to all hyperbolic domains $\Omega \subset \overline{\mathbb{C}}$.

Firstly, we cite some known results. In [121], Osgood proved that

$$\frac{1}{\alpha(\Omega)} \le \gamma(\Omega) \le \frac{2}{\alpha(\Omega)}, \quad \Omega \subset \mathbb{C}. \tag{3.37}$$

From Theorem 3.12 of Beardon and Pommerenke, it follows that

$$\frac{1}{\alpha(\Omega)} \le 2\pi M_0(\Omega) + \frac{(\Gamma(1/4))^4}{2\pi^2}, \quad \Omega \subset \mathbb{C}.$$

In [37], the first paper on uniformly perfect sets, it is also proved that

$$M_0(\Omega) \leq \frac{1}{2\alpha(\Omega)}, \quad \Omega \subset \mathbb{C}. \tag{3.38}$$

The following theorem assures that the equations (3.34) and (3.35) hold, where (3.34) is the sharp form of (3.38).

Theorem 3.20 ([27]). *Let $\Omega \subset \mathbb{C}$ be an open set with more than one boundary point in $\mathbb{C}$. If $\alpha(\Omega) > 0$, then the inequalities*

$$M_0(\Omega) < \frac{\gamma(\Omega)}{4} \quad and \quad M_0(\Omega) < \frac{1}{4\alpha(\Omega)} \tag{3.39}$$

are valid. The constant $1/4$ is best possible in both cases because of the equations

$$\lim_{\epsilon \to 0+} \frac{M_0(An(\epsilon, 1; 0))}{\gamma(An(\epsilon, 1; 0))} = \lim_{\epsilon \to 0+} M_0(An(\epsilon, 1; 0))\alpha(An(\epsilon, 1; 0)) = \frac{1}{4}. \tag{3.40}$$

Proof. We first prove (3.39). To that end, it is sufficient to consider the cases where $0 < M_0(\Omega) < \infty$, $\gamma(\Omega) < \infty$, and $\alpha(\Omega) > 0$. We have to estimate the modulus $M(An)$ for any genuine annulus $An := An(r_1, r_2; a) \subset \Omega$ such that

(i) $0 < r_1 < r_2 < \infty$,

(ii) $a \in \partial\Omega$,

(iii) $\{z \mid |z - a| = r_1\} \cup \partial\Omega \neq \emptyset$,

(iv) $\{z \mid |z - a| = r_2\} \cup \partial\Omega \neq \emptyset$.

We choose a point $b \in \partial\Omega$ with the property $|a - b| = r_1$ and define $z_0 \in An$ by

$$z_0 = a + \frac{b - a}{|b - a|} r_0, \quad r_0 = \sqrt{r_1 r_2}.$$

We have

$$\text{dist}(z_0, \partial\Omega) \leq r_0 - r_1 = r_0 \left(1 - e^{-\pi M(An)}\right)$$

and as in [37], p. 479, we get using the comparison principle (see [3], p. 53) for $An \subset \Omega$,

$$\lambda_\Omega(z_0) \leq \lambda_{An}(z_0) = \frac{1}{4r_0 M(An)}.$$

The two last formulas imply

$$\alpha(\Omega) \leq \lambda_\Omega(z_0)\text{dist}(z_0, \partial\Omega) \leq \frac{1 - e^{-\pi M(An)}}{4M(An)}.$$

Using the fact that $M_0(\Omega)$ is defined as the supremum of the moduli for the genuine annuli An defined as above and the first inequality in (3.37), we obtain

$$\frac{M_0(\Omega)}{1 - e^{-\pi M_0(\Omega)}} \leq \frac{1}{4\alpha(\Omega)} \leq \frac{\gamma(\Omega)}{4}. \tag{3.41}$$

Obviously, (3.41) implies (3.39).

As preparation for the proof of the equations (3.40), we first derive an interesting formula relating $M(An)$ and $\gamma(An)$ for genuine annuli An.

Lemma 3.21 ([27]). *For any genuine annulus An the equation*

$$\frac{(\gamma(An))^2}{16} = \frac{1}{4} + (M(An))^2 \tag{3.42}$$

is valid.

Proof. Since the conformal modulus and $|\nabla(1/\lambda_{An}(z))|$ are invariant under linear transformations of open sets, it is sufficient to consider annuli $An = An(\epsilon, 1; 0), \epsilon > 0$. Denoting by M the modulus of such an annulus, i.e.,

$$M := M(An) = \frac{1}{2\pi} \log(1/\epsilon),$$

we get

$$\frac{1}{\lambda_{An}(z)} = 4 M |z| \sin\left(\frac{\log(1/|z|)}{2 M}\right), \quad z \in An,$$

(see, for instance, [37]). Hence,

$$|\nabla(1/\lambda_{An}(z))| = \left|\frac{d(1/\lambda_{An}(z))}{d|z|}\right| =: |s(t)|,$$

where

$$s(t) = 4 M \sin t - 2 \sin t, \quad \text{with} \quad t = \frac{\log(1/|z|)}{2 M} \in (0, \pi).$$

By a straightforward calculation we derive

$$\gamma(An) = \max\{|s(t)| \mid t \in [0, \pi]\} = |s(t_0)| = 2\sqrt{1 + 4M^2}, \quad \tan t_0 = -2M,$$

which completes the proof of Lemma 3.21. $\square$

The formulas (3.39), (3.42) and Theorem 3.17 immediately imply

$$\lim_{\epsilon \to 0+} \frac{M_0(An(\epsilon, 1; 0))}{\gamma(An(\epsilon, 1; 0))} = \frac{1}{4}. \tag{3.43}$$

From the second inequality of (3.39) we see that

$$\overline{\lim}_{\epsilon \to 0+} M_0(An(\epsilon, 1; 0))\alpha(An(\epsilon, 1; 0)) \leq \frac{1}{4}.$$

Now, let us assume that the second equation in (3.40) is not true, i.e.,

$$\underline{\lim}_{\epsilon \to 0+} M_0(An(\epsilon, 1; 0))\alpha(An(\epsilon, 1; 0)) < \frac{1}{4}.$$

Applying once more Osgood's estimate $\alpha(\Omega) \geq 1/\gamma(\Omega)$ from (3.37) to $\Omega = An$ in the assumed inequality, we get

$$\underline{\lim}_{\epsilon \to 0+} \frac{M_0(An(\epsilon, 1; 0))}{\gamma(An(\epsilon, 1; 0))} < \frac{1}{4},$$

which contradicts (3.43).

The proof of Theorem 3.20 is complete. □

3.6 Derivatives of the conformal radius

There are two motivations to consider the derivatives of the conformal radius. The first one is the possibility to understand famous classical results in terms of the gradient of R. The classical inequalities on the second coefficient a_2 by Bieberbach [40] in the class S and by Löwner [109] in the family of convex univalent functions may serve as examples. Actually, let $z_0 \in \Omega$, and let f be a conformal map of the unit disc Δ onto $\Omega \subset \overline{\mathbb{C}}$ such that $f(\zeta_0) = z_0$, where $\zeta_0 \in \Delta$. Consider the Koebe transform

$$g(\zeta) = \frac{f\left(\frac{\zeta+\zeta_0}{1+\overline{\zeta_0}\zeta}\right) - f(\zeta_0)}{f'(\zeta_0)(1 - |\zeta_0|^2)} = \zeta + \sum_{n=2}^{\infty} a_n(g)\zeta^n, \quad \zeta \in \Delta.$$

It is well known that

$$2|a_2(g)| = \left|(1 - |\zeta_0|^2)\frac{f''(\zeta_0)}{f'(\zeta_0)} - 2\overline{\zeta_0}\right|.$$

On the other hand, by formula (3.6) one has

$$|\nabla R(z, \Omega)| = \left|(1 - |\zeta|^2)\frac{f''(\zeta)}{f'(\zeta)} - 2\overline{\zeta}\right| \tag{3.44}$$

for any $z = f(\zeta) \in \Omega$. Hence, Bieberbach's inequality $|a_2(g)| \leq 2$ immediately gives the following theorem.

Theorem 3.22 (Bieberbach). *If Ω is a simply connected proper subdomain of the plane $\mathbb{C}$, then*

$$|\nabla R(z, \Omega)| \leq 4, \quad z \in \Omega.$$

Clearly, $f(\Delta)$ is a convex domain if and only if $g(\Delta)$ is convex. Löwner's inequality $|a_2(g)| \leq 1$ implies that $|\nabla R(z_0, \Omega)| \leq 2$.

On the other hand, if $|\nabla R(z,\Omega)| \leq 2$ for any $z \in \Omega$, then (3.44) implies that

$$\operatorname{Re}\zeta \frac{f''(\zeta)}{f'(\zeta)} + 1 > 0, \quad \zeta \in \Delta,$$

which is the classical condition for a conformal map $f : \Delta \to \Omega$ to have a convex image. These give the following theorem.

Theorem 3.23 (Löwner). *A proper subdomain Ω of the plane $\mathbb{C}$ is convex if and only if*

$$\sup_{z \in \Omega} |\nabla R(z,\Omega)| \leq 2.$$

We have proved the following analog of Theorem 3.23.

Theorem 3.24 ([15], [20]). *Let $\Omega \subset \overline{\mathbb{C}}$ be a simply connected domain with more than one boundary point. The set $E = \mathbb{C} \setminus \Omega$ is convex if and only if*

$$\inf_{z \in \Omega} |\nabla R(z,\Omega)| \geq 2.$$

The second motivation is given by applications of the gradients to Schwarz-Pick type inequalities (see Chapter 4, Section 6, below) and by a number of theorems and their applications on the geometry of the surface

$$S_\Omega = \{(w,h) \mid w \in \Omega, h = R(w,\Omega)\}$$

(see [49], [74], [76], [89], [95], [119], [168], [169]). As an example we mention

Theorem 3.25. *A domain $\Omega \subset \mathbb{C}$ is convex if and only if $R(\cdot,\Omega)$ is a concave function on Ω.*

In the paper [95], Kovalev studied an analog of Theorem 3.25 for unbounded simply connected domains $\Omega \subset \overline{\mathbb{C}}$. He proved that $\mathbb{C} \setminus \Omega$ is a convex set if and only if the function $R(\cdot,\Omega)$ is a locally convex function. In [20] we developed these facts using the following observation.

Since the Jacobian of the gradient of R is

$$J(z,\Omega) = \frac{\partial^2 R(z,\Omega)}{\partial x^2} \frac{\partial^2 R(z,\Omega)}{\partial y^2} - \left(\frac{\partial^2 R(z,\Omega)}{\partial x \partial y}\right)^2, \quad z = x + iy \in \Omega,$$

and a condition necessary for a real-analytic function to be convex or concave on Ω is the inequality

$$J(z,\Omega) \geq 0, \quad z \in \Omega,$$

we observe that Theorem 3.25 and its generalizations are related to ∇R. Moreover, we consider the image of the domain Ω by the gradient (see [20] for details). We present here the case only when Ω is a polygonal domain.

Using the Wirtinger calculus, the gradient can be written as a complex variable function

$$\nabla R(z,\Omega) = \frac{\partial\, R(z,\Omega)}{\partial x} + i\frac{\partial\, R(z,\Omega)}{\partial y} = 2\,\frac{\partial\, R(w,\Omega)}{\partial\overline{z}},\ z = x + iy \in \Omega.$$

We begin with a simple example. Let $z = f(\zeta) = \zeta + 1/\zeta, \zeta \in \Delta$. One has $\Omega := f(\Delta) = \mathbb{C} \setminus [-2,2]$ and $\nabla R(\cdot,\Omega))$ is a diffeomorphism. Therefore, to find the gradient image it is sufficient to find its boundary. We have

$$g(\zeta) = \nabla R\left(\zeta + \frac{1}{\zeta},\Omega\right) = 2\,|1 - \zeta^2|\,\frac{1 - \zeta\overline{\zeta}^3}{\zeta(1 - \overline{\zeta}^2)^2},\ \zeta \in \Delta.$$

Hence,

$$\lim_{\zeta \to e^{i\theta}} g(\zeta) = \begin{cases} -2i & \text{for any } \theta \in (0,\pi), \\ 2i & \text{for any } \theta \in (\pi, 2\pi). \end{cases}$$

Moreover, it is clear that $g(\zeta)$ has no limit as $\zeta \to \pm 1$. Straightforward computations show that the set of all limit values of $g(\zeta)$ as $\zeta \to 1$, $\zeta \in \Delta$, is a curve γ_1 given by the parametric equation

$$w_1(t) = 2\,e^{it}\left(2\cos\frac{t}{2} - i\sin\frac{t}{2}\right),\ t \in (-\pi,\pi).$$

This is one branch between the two contact points $2i$ and $-2i$ of an epicycloid that is described by a point on a circle of radius 2 rolling on another circle of radius 2. Since $g(\zeta) = -g(-\zeta)$, the gradient image of Ω is the set of all points outside the two branches of the above epicycloid, where the second branch is described by

$$w_2(t) = -w_1(\pi + t),\ t \in (-\pi,\pi).$$

One may observe that γ_1 coincides with the set of values of ∇R for an angular domain with opening angle 2π, which is a branch of an epicycloid. This observation can be extended: If Ω is a polygonal domain, then the gradient image is bounded by branches of epicycloids or hypocycloids. To avoid confusion we want to mention that these epicycloids differ from those occurring in the Ptolemaic system.

Theorem 3.26 ([20]). *Let Ω be a simply connected domain in $\mathbb{C}$ or in $\overline{\mathbb{C}}$. If the boundary of Ω is a polygon with inner angles $\pi\alpha_k$, vertices z_k and sides (z_k, z_{k+1}), $k = 1,\ldots,n$, $z_{n+1} = z_1$, then*

(a) *$\nabla R(\cdot,\Omega)$ is a real-analytic function on $\overline{\Omega} \setminus \{z_1,\ldots,z_n,\infty\}$,*

(b) *the gradient image of a side (z_k, z_{k+1}) is a point w_k such that $|w_k| = 2$,*

(c) *the set of all limit values of ∇R as $z \to z_k$ is a curve γ_k that joins w_{k-1} with w_k and is defined by a parametric equation*

$$w_k(t) = 2\,e^{i(c_k + \alpha_k t)}(\alpha_k \cos t - i\sin t),\ t \in [-\frac{\pi}{2},\frac{\pi}{2}],$$

where c_k is a real constant.

Using Theorem 2.14 we shall obtain sharp estimates for higher-order derivatives of the conformal radius.. Let us introduce the functions

$$\mu_k(z, \Omega) := \frac{1}{(k-1)!} R(z, \Omega)^k \frac{\partial^k \log R(z, \Omega)^{-2}}{\partial z^k}, \quad k = 1, \ldots, n, \quad z \in \Omega.$$

Also, for a fixed $a \in \Omega$ we consider $\mu_k = \mu_k(a, \Omega)$ and the quantities

$$\tau_{n,n-1}(\alpha), \tau_{n,n-2}(\alpha), \ldots, \tau_{n,0}(\alpha)),$$

defined by the following recurrent formulas:

$$\tau_{k,k}(\alpha) = 1 \ (0 \le k \le n), \ \tau_{k,0}(\alpha) = \frac{\alpha}{k} \sum_{s=0}^{k-1} \mu_{k-s} \tau_{s,0}(\alpha), \ 1 \le k \le n, \tag{3.45}$$

$$\tau_{m,k}(\alpha) = \sum_{s=1}^{m-k+1} \frac{1}{s} \tau_{s-1,0}(1) \tau_{m-s,k-1}(\alpha), \quad 2 \le k \le m \le n. \tag{3.46}$$

In the case $0 \le k \le n - 1$, $\tau_{n,k}(\alpha)$ depends on n, k, α, a and Ω. For instance, if $\Omega = \Delta$ and $a \in \Delta$, then

$$\tau_{n,k}(\alpha) = \binom{n + 2\alpha - 1}{n - k} \bar{a}^{n-k}.$$

Theorem 3.27 ([14]). *If Ω is a simply connected proper subdomain of $\mathbb{C}$, then*

$$\sup_{\Omega} \sup_{a \in \Omega} |\mu_k(a, \Omega)| = \mu_k(a, \mathbb{C} \setminus [\frac{1}{4}, \infty)) \tag{3.47}$$

and

$$\sup_{\Omega} \sup_{a \in \Omega} |\tau_{n,k}(\alpha)| = \sum_{s=0}^{n-k} \binom{2n + 3\alpha - 1}{s} \binom{n - k - s + \alpha - 2}{n - k - s}. \tag{3.48}$$

Proof. Let ψ be the conformal mapping of Ω onto Δ, $\psi(a) = 0$, $\psi'(a) = \lambda_\Omega(a) > 0$. By definition,

$$\lambda_\Omega(z) = |\psi'(z)|(1 - |\psi(z)|^2)^{-1}, \quad z \in \Omega.$$

This yields

$$\frac{\partial \log \lambda_\Omega^2(z)}{\partial z} = \frac{\psi''(z)}{\psi'(z)} - \frac{\overline{2\psi(z)}}{1 - \psi(z)\overline{\psi(z)}} \psi'(z).$$

Hence

$$\lambda_\Omega^k(a)\mu_k(a, \Omega) = \frac{1}{(k-1)!} \frac{\partial^k \log \lambda_\Omega^2(a)}{\partial a^k} = \frac{1}{(k-1)!} \left(\frac{\psi''(z)}{\psi'(z)} \right)^{(k-1)} \big|_{z=a} \tag{3.49}$$

and

$$\frac{\psi''(z)}{\psi'(z)} = \sum_{k=0}^{\infty} \lambda_{\Omega}^{k+1}(a) \mu_{k+1}(a, \Omega)(z-a)^k \tag{3.50}$$

in some neighbourhood of a.

According to Theorem 2.14 by Klouth and Wirths and the equation (3.50),

$$|\mu_k(a, \Omega)| \leq \mu_k(0, \mathbb{C} \setminus (\frac{1}{4}, \infty)), \quad k = 1, 2, \ldots.$$

Equality for $k \in \mathbb{N}$ occurs if and only if $\{a, \Omega\}$ is a linear transformation of $\{0, \mathbb{C} \setminus [\frac{1}{4}, \infty)\}$. The conformal mapping $\Phi_0 : \Delta \to \mathbb{C} \setminus [\frac{1}{4}, \infty)$, $\Phi_0(0) = 0$, $\Phi_0(0) > 0$, is given by the Koebe function $\Phi_0(\zeta) = \zeta(1+\zeta)^{-2}$. It is not difficult to verify by direct calculations that

$$\tau_{n,k}^0(\alpha) = \sum_{s=0}^{n-k} \binom{2n+3\alpha-1}{s} \binom{n-k-s+\alpha-2}{n-k-s}$$

for the Koebe domain $\mathbb{C} \setminus [\frac{1}{4}, \infty)$ at the point $z = 0$.

From (3.47) we obtain (3.48) since $\tau_{n,k}(\alpha)$ is a polynomial with positive coefficients in the $\mu_1, \ldots, \mu_n$.

For $n = 1$, we get $|\tau_{1,0}(\alpha)| = \alpha|\text{grad}\,\lambda_{\Omega}^{-1}(a)|$. Hence, $|\tau_{1,0}(\alpha)| \leq 4\alpha$ by the classical Koebe constant $1/4$. $\square$

Chapter 4

Basic Schwarz-Pick type inequalities

Let $\Omega \subset \overline{\mathbb{C}}$ and $\Pi \subset \overline{\mathbb{C}}$ be two domains equipped by the Poincaré metric. We are concerned with the set

$$A(\Omega, \Pi) = \{f : \Omega \to \Pi\}$$

of functions locally holomorphic or meromorphic in Ω and, in general, multivalued. Let $\lambda_\Omega(z), z \in \Omega$, and $\lambda_\Pi(w), w \in \Pi$, denote the density of the Poincaré metric at $z \in \Omega$ and $w \in \Pi$, respectively.

Consider the functional $L_n(f, z, \Omega, \Pi)$ defined by

$$\frac{\left|f^{(n)}(z)\right|}{n!} = L_n(f, z, \Omega, \Pi) \frac{(\lambda_\Omega(z))^n}{\lambda_\Pi(f(z))}, \quad z \in \Omega, \quad f \in A(\Omega, \Pi), \quad n \in \mathbb{N}.$$

Many problems in geometric function theory are devoted to the problem of determining

$$M_n(z, \Omega, \Pi) := \sup\{L_n(f, z, \Omega, \Pi) \mid f \in A(\Omega, \Pi)\},$$

and, respectively,

$$C_n(\Omega, \Pi) := \sup\{M_n(z, \Omega, \Pi) \mid z \in \Omega\}.$$

It is clear that $C_n(\Omega, \Pi)$ is not dependent on f and $z \in \Omega$ and represents the smallest number possible in the inequality

$$\frac{\left|f^{(n)}(z)\right|}{n!} \leq C_n(\Omega, \Pi) \frac{(\lambda_\Omega(z))^n}{\lambda_\Pi(f(z))}, \quad z \in \Omega, \quad f \in A(\Omega, \Pi).$$

The classical Schwarz-Pick lemma says that $M_1(z, \Delta, \Delta) = C_1(\Delta, \Delta) = 1$ and in turn $M_1(z, \Omega, \Pi) = C_1(\Omega, \Pi) = 1$ for any pair (Ω, Π) of hyperbolic domains in the extended complex plane, as we have discussed above.

We shall consider the problem of determining $C_n(\Omega, \Pi)$ for all $n \geq 2$. In the proofs we will frequently use the fact that the functions M_n and C_n are invariant under linear transformation of domains. This means that the equations

$$M_n(z, \Omega, \Pi) = M_n(az + b, a\Omega + b, c\Pi + d)$$

and

$$C_n(\Omega, \Pi) = C_n(a\Omega + b, c\Pi + d)$$

are valid, where $a\Omega + b = \{az + b \mid z \in \Omega\}$ and $c\Pi + d = \{cw + d \mid w \in \Pi\}$ for some $a, c \in \mathbb{C} \setminus \{0\}$, $b, d \in \mathbb{C}$.

Also, the following fact deserves reader's attention. For technical reasons we explain a simple case, when Π is a bounded domain and $z \in \Omega \subset \mathbb{C}$. Normal family arguments show that there exists a sequence $f_k \in A(\Omega, \Pi)$ such that

$$M_n(z, \Omega, \Pi) = \lim_{k \to \infty} L_n(f_k, z, \Omega, \Pi),$$

and f_k converges to a holomorphic function f_0 uniformly in the interior of the domain Ω. Clearly, there are two possible cases to distinguish: The limit function f_0 belongs to the family $A(\Omega, \Pi)$ or $f_0(z) \equiv const. \in \partial\Pi$, and consequently, $f_0 \notin A(\Omega, \Pi)$. The second case is typical in our problems. If $n \geq 2$, then there is no extremal function $f \in A(\Omega, \Pi)$ such that $M_n(z, \Omega, \Pi) = L_n(f, z, \Omega, \Pi)$ except in some very special cases.

4.1 Two classical inequalities

Let $f \in A(\Delta, \Delta)$. A consequence of

$$|f'(z_0)| \leq \frac{1 - |f(z_0)|^2}{1 - |z_0|^2}, \quad z_0 \in \Delta,$$

is the sharp inequality

$$|f'(z_0)| \leq \frac{1}{1 - |z_0|^2}, \quad z_0 \in \Delta.$$

Equality is attained for conformal automorphisms f of Δ such that $f(z_0) = 0$. In 1920, Szász extended the latter inequality to higher-order derivatives.

Theorem 4.1 (O. Szász [160]). *For any $f \in A(\Delta, \Delta)$, $z \in \Delta$, and $m \in \mathbb{N}$, the sharp inequality*

$$\left| f^{(2m+1)}(z) \right| \leq \frac{(2m+1)!}{(1 - |z|^2)^{2m+1}} \sum_{k=0}^{m} \binom{m}{k}^2 |z|^{2k}$$

is valid. Equality occurs only for the functions

$$f(\zeta) = e^{i\gamma} \zeta^m \left(\frac{\zeta - z}{1 - \bar{z}\zeta} \right)^{m+1}, \quad \gamma \in \mathbb{R}.$$

Proof. We consider the function $g \in A(\Delta, \Delta)$ defined by

$$g(\zeta) = f\left(\frac{\zeta + z}{1 + \overline{z}\zeta}\right) = \sum_{k=0}^{\infty} a_k \zeta^k, \quad \zeta \in \Delta.$$

According to the Cauchy formula, integration along a circle Γ around the origin lying in its neighbourhood and the use of the variable substitution

$$\xi = \frac{\zeta + z}{1 + \overline{z}\zeta} = S(\zeta)$$

results in the following chain of equations:

$$\frac{f^{(n)}(z)}{n!} = \frac{1}{2\pi i} \int_{S(\Gamma)} \frac{f(\xi)\,d\xi}{(\xi - z)^{n+1}} = \frac{1}{2\pi i} \int_{\Gamma} \frac{g(\zeta)(1 + \overline{z}\zeta)^{n-1}d\zeta}{\zeta^{n+1}(1 - |z|^2)^n}. \tag{4.1}$$

For $n = 2m + 1$ using $|g(\zeta)| \leq 1$ we obtain

$$\frac{1}{2\pi}\left|\int_{\Gamma} \frac{g(\zeta)(1 + \overline{z}\zeta)^{2m}d\zeta}{\zeta^{2m+2}}\right| \leq \frac{1}{2\pi}\int_0^{2\pi} |1 + \overline{z}e^{i\theta}|^{2m}d\theta. \tag{4.2}$$

Clearly, formulas (4.1) and (4.2) imply the desired estimate by the binomial and Parseval formulas. A little examination of the inequality (4.2) gives that equality can occur only for the function

$$g(\zeta) = e^{i\gamma}\zeta^{m+1}\left(\frac{\zeta + z}{1 + \overline{z}\zeta}\right)^m, \quad \gamma \in \mathbb{R}.$$

This completes the proof of Theorem 4.1. $\qquad\square$

Remark 4.2. In [160], during the proof of the theorem Szász gave the formula

$$\frac{(1 - |z|^2)^n}{n!}f^{(n)}(z) = \sum_{k=1}^{n} \binom{n-1}{n-k} a_k \overline{z}^{n-k} \tag{4.3}$$

as a consequence of equation (4.1). Also, in [160] by a similar proof Szász obtained an explicit sharp bound for $|f''(z)|$. But the problem of finding an explicit formula for $\{\max |f^{(2m)}(z)| \mid f \in A(\Delta, \Delta)\}$, $z \in \Delta$, in the case $m \geq 2$ is still open (2008). Concerning Szász's majorant for $n = 2m + 1$, one easily obtains that

$$\sup_{z \in \Delta} \frac{1}{2\pi}\int_0^{2\pi} |1 + \overline{z}e^{i\theta}|^{2m}d\theta = \frac{1}{2\pi}\int_0^{2\pi} |1 + e^{i\theta}|^{2m}d\theta = 2^{2m}\frac{(2m-1)!!}{(2m)!!}.$$

Moreover, for $n = 2m$ the proof gives the estimate

$$\sup_{z \in \Delta} \frac{(1 - |z|^2)^{2m}}{(2m)!}|f^{(2m)}(z)| \leq \frac{2^{2m}}{\pi}\frac{(2m-2)!!}{(2m-1)!!}$$

which is not sharp, at least for $n = 2$.

E. Landau remarked in [98] that a special case of the following theorem, namely, Theorem 7.1 (see below), is a consequence of the validity of the Bieberbach conjecture.

Theorem 4.3. *Let Π be a simply connected proper subdomain of $\mathbb{C}$ and $z \in \Delta$. Then*

$$M_n(z, \Delta, \Pi) \leq M_n(z, \Delta, H_2) = (n + |z|)(1 + |z|)^{n-2},$$

where $H_2 = \mathbb{C} \setminus (-\infty, 0]$.

Proof. We use the fact that de Branges' proof of the Bieberbach conjecture implies a proof of the generalized Bieberbach or Rogosinski conjecture. This means that the Taylor coefficients of a function subordinate to a schlicht function are dominated by the Taylor coefficients of the Koebe function. Therefore, for the Taylor coefficients of a function $f \in A(\Delta, \Pi)$, the inequalities

$$\lambda_\Pi(a_0)\,|a_k| \leq k$$

are valid. Using the latter inequalities and formula (4.3), we get

$$M_n(z, \Delta, \Pi) \leq \sum_{k=1}^{n} \binom{n-1}{n-k} k|z|^{n-k} = (n + |z|)(1 + |z|)^{n-2}.$$

Straightforward computations show that

$$L_n(g_0, z, \Delta, H_2) = M_n(z, \Delta, H_2) = (n + |z|)(1 + |z|)^{n-2},$$

where

$$g_0(\zeta) = \frac{1}{4}\left(\left(\frac{1 + \zeta\,|z|/z}{1 - \zeta\,|z|/z}\right)^2 - 1\right) = \sum_{k=1}^{\infty} k(\zeta\,|z|/z)^k.$$

This completes the proof of Theorem 4.3.   □

Corollary 4.4. *If Π is a simply connected proper subdomain of $\mathbb{C}$, then*

$$C_n(\Delta, \Pi) \leq C_n(\Delta, H_2) = 2^{n-2}(n + 1).$$

4.2   Theorems of Ruscheweyh and Yamashita

According to our notation $C_n(\Omega, \Pi)$, the identity

$$C_n(\Delta, \Pi) = 2^{n-1}$$

has been proved by St. Ruscheweyh (see [142] and [143]) in two basic cases, when Π is a half plane or a disc. Here we present the original versions of his theorems.

Theorem 4.5 (St. Ruscheweyh [143] (1985)). *Let* $\Delta = \{\zeta \mid |\zeta| < 1\}$. *For any* $f \in A(\Delta, \Delta)$, $z \in \Delta$, *and* $n \in \mathbb{N}$ *the sharp inequality*

$$\frac{1}{n!}\left|f^{(n)}(z)\right| \leq \frac{1 - |f(z)|^2}{(1 - |z|)^n (1 + |z|)}$$

is valid.

This inequality was conjectured by Ruscheweyh in 1974 (see [142]).

Proof. Let the function

$$g(z) = \sum_{k=0}^{\infty} a_k z^k$$

be holomorphic and $|g(z)| \leq 1$ in the unit disc. It is well known that the coefficients of such a function satisfy the inequalities

$$|a_0| \leq 1, \quad |a_k| \leq 1 - |a_0|^2 \tag{4.4}$$

for any $k \geq 1$.

The inequalities (4.4) imply the assertion of the theorem immediately for $z = 0$. We now suppose that $z \neq 0$ and define $g(\zeta)$ as in the proof of Theorem 4.1.

Using the Szász formula (4.3) and (4.4) together with equations $f(z) = h(0) = b_0$, one easily gets

$$\frac{(1 - |z|^2)^n}{n!}\left|f^{(n)}(z)\right| \leq (1 - |b_0|^2) \sum_{k=1}^{n} \binom{n-1}{n-k} |z|^{n-k} = (1 - |b_0|^2)(1 + |z|)^{n-1}$$

which is the inequality to prove.

We have to show that the inequality of Theorem 4.5 is sharp for $n \geq 2$ and any $z \in \Delta$, $z \neq 0$. To this end we choose a sequence a_k in Δ such that $a_k \to z/|z|$ as $k \to \infty$ and consider unimodular bounded holomorphic functions

$$f_k(z) = \frac{z - a_k}{1 - \overline{a_k} z}.$$

Straightforward computations give

$$\frac{|f_k^{(n)}(z)|}{1 - |f_k(z)|^2} = \frac{n! |a_k|^{n-1}}{|1 - \overline{a_k} z|^{n-1}(1 - |z|^2)} \to \frac{n!}{(1 - |z|)^{n-1}(1 - |z|^2)}$$

as $k \to \infty$.

This completes the proof of Theorem 4.5. $\qquad\square$

Theorem 4.6 (St. Ruscheweyh [142]). *Let* f *be holomorphic in* Δ, $n \in \mathbb{N}$, *and* $\rho(f(z))$ *denote the minimal distance from* $f(z)$ *to the boundary of the closed convex hull of* $f(\Delta)$. *Then the sharp inequality*

$$\frac{1}{n!}\left|f^{(n)}(z)\right| \leq \frac{2\,\rho(f(z))}{(1 - |z|)^n(1 + |z|)}$$

is valid. Equality occurs for conformal maps of the unit disc onto a halfplane.

Remark 4.7. For $H_1 = \{z \mid \operatorname{Re} z > 0\}$ this implies that for any $f \in A(\Delta, H_1), z \in \Delta$ and $n \in \mathbf{N}$ the inequality

$$\frac{1}{n!} \left| f^{(n)}(z) \right| \le (1 + |z|)^{n-1} \frac{(\lambda_\Delta(z))^n}{\lambda_{H_1}(f(z))}$$

holds.

Proof of Theorem 4.6. (As St. Ruscheweyh indicated in [142], he owes the idea of the following proof to T. Sheil-Small.) Without loss of generality we can suppose that $\operatorname{Re} f(\zeta) > 0$ in Δ, $f(0) = 1$ and $\rho(f(z)) = \operatorname{Re} f(z)$ for a given point $z \in \Delta$. Using the Herglotz formula

$$f(\zeta) = \int_0^{2\pi} \frac{1 + \zeta e^{-it}}{1 - \zeta e^{-it}} d\mu(t),$$

where $\mu(t)$ is a nondecreasing function with $\mu(2\pi) - \mu(0) = 1$, we get

$$f^{(n)}(\zeta) = 2n! \int_0^{2\pi} \frac{e^{-int}}{(1 - \zeta e^{-it})^{n+1}} d\mu(t).$$

Consequently, by simple estimates and the Poisson formula

$$\left| f^{(n)}(z) \right| \le \frac{2n!}{(1 - |z|)^{n-1}(1 - |z|^2)} \int_0^{2\pi} \frac{1 - |z|^2}{|1 - ze^{-it}|^2} d\mu(t) = \frac{2n! \operatorname{Re} f(z)}{(1 - |z|)^{n-1}(1 - |z|^2)}.$$

Since $1 - |z| = |1 - ze^{-it}|$ is true only for $e^{-it} = e^{-it_0} = \overline{z}/|z|$, it is easily seen that equality can occur only for a piece-wise constant function $\mu(t)$ such that $\mu([0, 2\pi]) = \{0, 1\}$, and, consequently, the corresponding function f_0 has the form

$$f_0(\zeta) = \frac{1 + \zeta e^{-it_0}}{1 - \zeta e^{-it_0}}, \quad e^{it_0} = z/|z|,$$

and f_0 maps Δ onto H_1. This completes the proof of Theorem 4.6. $\qquad\square$

Let Ω be a hyperbolic domain in $\mathbb{C}$, i.e., Ω has more than two boundary points. For a universal covering map φ from Δ onto Ω and given $z \in \Omega$, let $\rho_\Omega(z)$ denote the greatest $r \in (0, 1]$ such that φ is univalent in the non-Euclidean disc

$$\left\{ \zeta \; \middle| \; \left| \frac{\zeta - w}{1 - \overline{w}\zeta} \right| < r \right\}$$

where $z = \varphi(w)$. Since the hyperbolic distance is a conformal invariant, the radius of univalence $\rho_\Omega(z)$ is well defined. Also, we need the notation

$$H_2 = \mathbb{C} \setminus (-\infty, 0], \quad H_1 = \{w \mid \operatorname{Re} w > 0\}.$$

Theorem 4.8 (S. Yamashita [170]). *Let Ω be a hyperbolic domain in $\mathbb{C}$, let $z \in \Omega$, and let $n \geq 2$. For any holomorphic function $f : \Omega \to \mathbb{C}$ with positive real part in Ω the sharp inequality*

$$\frac{1}{n!} \left| f^{(n)}(z) \right| \leq 2 \binom{2n-1}{n} \left(\frac{\lambda_\Omega(z)}{\rho_\Omega(z)} \right)^n \operatorname{Re}(f(z))$$

is valid.

Also, in [170] Yamashita established the case of equality in Theorem 4.8. In particular, if $\Omega = H_2$ and $f(\zeta) = \zeta^{-1/2}$, then the equality is attained at the points $\zeta = z = x > 0$. Taking into account the classical equality $1/\lambda_{H_1}(w) = 2\operatorname{Re} w$ and the invariance of C_n under linear transformation of domains, one obtains the equality

$$C_n(aH_2 + b, H_1) = \binom{2n-1}{n} \quad (a \neq 0).$$

Clearly, if Ω is a simply connected domain, then $\rho_\Omega(z) = 1$ for any point $z \in \Omega$. Therefore, this particular case of Yamashita's theorem can be presented as follows.

Theorem 4.9. *If Ω is a simply connected proper subdomain of $\mathbb{C}$, then*

$$C_n(\Omega, H_1) \leq C_n(H_2, H_1) = \binom{2n-1}{n}.$$

In fact, Theorem 4.9 is equivalent to Theorem 4.8. Actually, suppose that Ω is a multiply connected domain in $\mathbb{C}$ and φ as above. The covering map $\varphi_1 : \Delta \to \Omega$ defined by

$$\varphi_1(\zeta) = \varphi\left(\frac{\zeta + w}{1 + \overline{w}\zeta} \right)$$

is univalent in the disc $\Delta_\rho = \{\zeta \,|\, |\zeta| < \rho_\Omega(z)\}$. The domain $\Omega_\rho := \varphi(\Delta_\rho)$ is a simply connected proper subdomain of $\mathbb{C}$ and

$$\frac{\lambda_\Omega(z)}{\rho_\Omega(z)} = \lambda_{\Omega_\rho}(z).$$

Applying Theorem 4.9 to the function $f \,|\, \Omega_\rho$ in the simply connected domain Ω_ρ one immediately obtains Theorem 4.8.

The original proof of Yamasita of these theorems is based on coefficient estimates of univalent functions by Chua (see [56] and [170]). We will obtain Theorem 4.9 as a special case of our theorem proved in Section 4 of the present chapter.

4.3 Pairs of simply connected domains

We again will need the domain $H_2 = \mathbb{C} \setminus (-\infty, 0]$. In this section we will discuss the following theorem.

Theorem 4.10 ([16]). *If Ω and Π are simply connected proper subdomains of $\mathbb{C}$, then the sharp estimate*

$$C_n(\Omega, \Pi) \leq C_n(H_2, H_2) = 4^{n-1}$$

is valid.

Proof. For $f \in A(\Omega, \Pi)$, $z_0 \in \Omega$, we consider the functions

$$s(\zeta) := (\Phi_{\Omega, z_0}(\zeta) - z_0)\, \lambda_\Omega(z_0), \quad \zeta \in \Delta,$$

and

$$t(\zeta) := \left(\Phi_{\Pi, f(z_0)}(\zeta) - f(z_0)\right) \lambda_\Pi(f(z_0)), \quad \zeta \in \Delta.$$

Both of them belong to the class S of functions univalent in Δ and normalized in the origin as usual. The fact that $f(\Omega)$ is a subset of Π may be expressed in terms of the function

$$u(\zeta) := (f(\Phi_{\Omega, z_0}(\zeta)) - f(z_0))\, \lambda_\Pi(f(z_0)), \quad \zeta \in \Delta.$$

It means that $u(\zeta)$ is subordinate to $t(\zeta)$.

Using the Taylor expansion

$$u(\zeta) = \sum_{k=1}^{\infty} a_k \lambda_\Pi(f(z_0)) \zeta^k$$

and the function Ψ_{Ω, z_0} inverse to Φ_{Ω, z_0} we get

$$f(z) = f(z_0) + \sum_{k=1}^{\infty} a_k \left(\Psi_{\Omega, z_0}(z)\right)^k$$

and therefore

$$\frac{f^{(n)}(z_0)}{n!} = \sum_{k=1}^{n} a_k \frac{1}{n!} \left.\left(\left(\Psi_{\Omega, z_0}(z)\right)^k\right)^{(n)}\right|_{z=z_0}.$$

If we denote by $s^{-1}(w)$ the function inverse to $s(\zeta)$, define

$$\left(s^{-1}(w)\right)^k = \sum_{m=k}^{\infty} A_{m,k}(z_0) w^m$$

and use

$$s^{-1}\left(\lambda_\Omega(z_0)(z - z_0)\right) = \Psi_{\Omega, z_0}(z),$$

we see that

$$\left.\frac{1}{n!} \left(\left(\Psi_{\Omega, z_0}(z)\right)^k\right)^{(n)}\right|_{z=z_0} = A_{n,k}(z_0)\, (\lambda_\Omega(z_0))^n.$$

Hence, we get the formula

$$\frac{f^{(n)}(z_0)}{n!} = \sum_{k=1}^{n} a_k \, A_{n,k}(z_0) \, (\lambda_\Omega(z_0))^n \,, \qquad (4.5)$$

which will be central in what follows.

Let K_1 be the inverse of the Koebe function

$$k_1(z) = \frac{z}{(1+z)^2}.$$

We have

$$K_1(w) = w + \sum_{m-2}^{\infty} A_m(K_1) \, w^m, \quad A_m(K_1) = \frac{(2m)!}{m!(m+1)!},$$

and, by the Löwner theorem 2.13, the sharp estimates

$$|A_{m,1}| \leq A_m(K_1) = \frac{(2m)!}{m!(m+1)!}$$

are valid. Let us define $A_{m,k}(K_1)$ by expansions

$$(K_1(w))^k = \sum_{m=k}^{\infty} A_{m,k}(K_1) \, w^m$$

in some neighbourhood of the origin. It is evident that $A_{n,k}(K_1)$ is a polynomial with positive coefficients in $A_m(K_1)$, $2 \leq m \leq n$, and $A_{n,k}(z_0)$ is the same polynomial in $A_{m,1}(z_0)$, $2 \leq m \leq n$. Accordingly, we have

$$|A_{n,k}(z_0)| \leq A_{n,k}(K_1), \quad 1 \leq k \leq n. \qquad (4.6)$$

For those quantities $A_{n,k}(K_1)$ we get

$$A_{n,k}(K_1) = \frac{k}{n} \begin{pmatrix} 2n \\ n-k \end{pmatrix} = \frac{2k(2n-1)!}{(n-k)!(n+k)!}. \qquad (4.7)$$

This is an immediate consequence of the Cauchy formula according to

$$A_{n,k}(K_1) = \frac{1}{2\pi i} \int_{k_1(\partial \Delta_r)} \frac{(K_1(w))^k}{w^{n+1}} \, dw = \frac{1}{2\pi i} \int_{\partial \Delta_r} \frac{(1+\zeta)^{2n-1}(1-\zeta)}{\zeta^{n-k+1}} \, d\zeta,$$

where $r \in (0,1)$ and $\partial \Delta = \{\zeta \mid |\zeta| = r\}$ (for (4.6) and (4.7) compare [56]).

Further, the function u is subordinate to the function t univalent in Δ and normalized as usual. According to the Rogosinski conjecture settled by de Branges'

proof of the Bieberbach conjecture, the Taylor coefficients of the function u satisfy
the inequalities

$$|a_k|\,\lambda_\Pi(f(z_0)) \le k, \quad k \in \mathbb{N}.$$

Using these estimates and the basic formula (4.5), we obtain

$$\frac{|f^{(n)}(z_0)|}{n!}\,\frac{\lambda_\Pi(f(z_0))}{(\lambda_\Omega(z_0))^n} \le \sum_{k=1}^{n} k\,A_{n,k}(K_1),$$

where $A_{n,k}(K_1)$ are given by formula (4.7). Using again the Cauchy formula to
compute the latter sum, we get

$$\sum_{k=1}^{n} \frac{k^2}{n}\binom{2n}{n-k} = 4^{n-1}$$

(for the latter formula see also [56]). To complete the proof we have to show that
$C_n(H_2, H_2) = 4^{n-1}$, where $H_2 = \mathbb{C}\setminus(-\infty, 0]$. To this end we consider $f_0(z) = 1/z$,
$z \in H_2$. For any $z = x > 0$ using $\lambda_{H_2}(x) = 1/(4x)$ and $\lambda_{H_2}(1/x) = x/4$ one easily
gets

$$\frac{|f_0^{(n)}(x)|}{n!}\,\frac{\lambda_{H_2}(1/x)}{(\lambda_{H_2}(x))^n} = 4^{n-1}.$$

This completes the proof of Theorem 4.10. $\square$

We will see below that it is possible to find lower bounds for punishing
factors $C_n(\Delta, \Pi)$ using the properties of hyperbolic metrics if the boundaries of
the domains in question are "nice" (see Section 5.5). This was the reason why
we tried to find lower bounds for punishing factors for pairs of simply connected
domains by considering the limiting processes going to the boundaries. Our result
is the following theorem.

Theorem 4.11 (see [16]). *Let Ω and Π be two simply connected proper subdomains
of $\mathbb{C}$ whose boundaries contain sectorial accessible analytic arcs. Then, for any
$n \ge 2$, the assertion*

$$2^{n-1} \le C_n(\Delta, \Pi) \le C_n(\Omega, \Pi)$$

holds. Equality occurs if $\Omega = \Delta$ and Π is convex.

As usual, a boundary arc is said to be sectorial accessible if any point on this
arc is the vertex of an open triangle contained in the domain in question (see for
instance [131]). The proof for this result is very technical. Therefore we omit the
details here. We hope that one of our readers will be able to give a proof for the
following conjecture.

Conjecture (see [16]). Given $n \ge 3$, then $C_n(\Omega, \Pi) \ge 2^{n-1}$ for all simply connected
domains Ω and Π in $\mathbb{C}$.

4.4 Holomorphic mappings into convex domains

We shall consider a holomorphic function $f : \Omega \to \Pi$, where Π is a proper convex subdomain of the plane and Ω is the unit disc or a simply connected domain.

The following theorem is a generalization of the Carathéodory inequality ([51], [52], see also [42], p. 365, [48], p. 213) for Taylor coefficients of holomorphic functions with positive real part. Also, in [83] Herzig gives a description of extremal functions.

Theorem 4.12 (see Rogosinski [139]). *Let g and g_0 be holomorphic functions with expansions*

$$g(z) = \sum_{n=1}^{\infty} a_n z^n, \qquad g_0(z) = z + \sum_{n=2}^{\infty} b_n z^n, \quad z \in \Delta.$$

Suppose that g_0 is univalent in Δ and $\Pi := g_0(\Delta)$ is a convex domain. If $g \prec g_0$, then $|a_n| \leq 1$ for any $n \geq 1$. Equalities $|a_k| = 1$ for all $k = 1, 2, \ldots, n$ with some $n \geq 2$ occur if and only if g is a conformal map of Δ onto a half plane.

Proof. For $n \in \mathbb{N}$ we consider the function

$$g_n(z) := \frac{1}{n} \sum_{k=1}^{n} g(\zeta^k z^{1/n}) = a_n z + \sum_{k=2}^{\infty} a_{nk} z^k, \quad z \in \Delta,$$

where $\zeta = e^{2\pi i/n}$. Since $w_k = g(\zeta^k z) \in \Pi$ for any $z \in \Delta$, from the convexity of Π it follows that $(w_1 + \cdots + w_n)/n \in \Pi$. Hence, the function g_n is subordinate to g_0. Applying the Schwarz lemma to the function $g_0^{-1}(g_n(z))$, one immediately gets the desired inequality $|a_n| \leq 1$ for any $n \geq 1$. If $|a_1| = 1$, then Schwarz's lemma implies that $g(z) = g_0(cz)$, where $|c| = 1$. If $|a_1| = |a_2| = 1$, then $g(z) = g_0(cz)$ and $|b_2| = 1$. Consequently, g is a conformal map of the unit disc onto a half plane according to the classical Löwner theorem on coefficients of convex univalent functions.

The proof is complete. $\qquad\square$

The assertion of the following theorem immediately implies that $M_n(\Delta, \Pi) = 2^{n-1}$ for any convex domain Π.

Theorem 4.13 ([16]). *Let Π be a convex proper subdomain of $\mathbb{C}$ and let $n \in \mathbb{N}$. Then for any $z \in \Delta$ the equation*

$$M_n(z, \Delta, \Pi) = (1 + |z|)^{n-1} \tag{4.8}$$

is valid. In the case $z \neq 0$ and $n \geq 2$, there exist extremal functions for which

$$L_n(f, z, \Delta, \Pi) = (1 + |z|)^{n-1}$$

if and only if Π is a half plane.

Proof. We fix $z \in \Delta$ and consider $f \in A(\Delta, \Pi)$. Introduce the function $f_1 \in A(\Delta, \Pi)$ defined by

$$f_1(\zeta) = f\left(\frac{\zeta + z}{1 + \overline{z}\zeta}\right) = \sum_{k=0}^{\infty} a_k \zeta^k, \quad \zeta \in \Delta.$$

Using the formula (4.3) we obtain

$$M_n(z, \Delta, \Pi) \leq \sup\left\{\sum_{k=1}^{n} \binom{n-1}{n-k} |z|^{n-k} |a_k| \lambda_\Pi(a_0) \mid g \in A(\Delta, \Pi)\right\}. \quad (4.9)$$

From Theorem 4.12 for the function $g(\zeta) = \lambda_\Pi(a_0) f_1(\zeta)$ it follows that

$$\lambda_\Pi(a_0)|a_k| \leq 1, \quad k \in \mathbb{N}.$$

This together with inequality (4.9) give

$$M_n(z, \Delta, \Pi) \leq \sum_{k=1}^{n} \binom{n-1}{n-k} |z|^{n-k} = (1 + |z|)^{n-1}.$$

To prove $M_n(z, \Delta, \Pi) \geq (1 + |z|)^{n-1}$ we first consider the case $z = 0$. If Φ denotes the conformal map of the unit disc Δ onto Π with $\Phi(0) = w_0$ and $\Phi'(0) = 1/\lambda_\Pi(w_0) > 0$, it is obvious that the function f defined by $f(\zeta) = \Phi(\zeta^n)$ has the desired property: $f^{(n)}(0)\lambda_\Pi(w_0)/n! = 1$.

The inequalities for coefficients in Theorem 4.12 occur for all $k = 1, 2, \ldots, n$ with $n \geq 2$ if and only if g is a conformal map of the unit disc onto a half plane. Therefore the existence of a function f for which $L_n(f, z, \Delta, \Pi) = (1 + |z|)^{n-1}$ in the case $z \neq 0$ is possible only if Π is a half plane and this is the case in Theorem 4.5 of Ruscheweyh.

In the general case we only know that there is a maximising sequence $(f_k) \subset A(\Omega, \Pi)$ such that

$$M_n(z, \Delta, \Pi) = \lim_{k \to \infty} L_n(f_k, z, \Delta, \Pi).$$

Without loss of generality, we can suppose that the sequence (f_k) converges uniformly in the interior of the unit disc. Let

$$f_0(z) = \lim_{k \to \infty} f_k(z), \quad z \in \Delta.$$

If $z \neq 0$, $n \geq 2$, and Π is not a half plane, then the proof gives that $f_0 \notin A(\Delta, \Pi)$. Hence, $f_0(z) \equiv const. \in \partial\Pi$. To prove, nevertheless, the sharpness of (4.8) for any convex domain Π, we present such a sequence explicitly using the convexity of Π and the linear invariance of $M_n(z, \Delta, \Pi)$.

Clearly, there exist a point $w_1 \in \partial\Pi$, a disc Δ_1 and a half plane H such that $\Delta_1 \subset \Pi \subset H$ and $w_1 \in \partial\Delta_1 \cap \partial\Pi \cap \partial H$. Without loss of generality we may suppose that

$$\Delta_1 = \{w \mid |w - 1| < 1\} \subset \Pi \subset H_1.$$

The origin belongs to the boundaries of Δ_1, Π and H_1. Now we use a refinement of an idea presented in the proof of Theorem 4.5 by Ruscheweyh. Namely, for fixed $z \in \Delta \setminus \{0\}$, we consider the sequence

$$\alpha_k = \left(1 - \frac{1}{k+1}\right)\frac{z}{|z|}, \quad k \in \mathbb{N},$$

of complex numbers and the sequence of f_k, $k \in \mathbb{N}$, of conformal maps of Δ onto Δ_1 defined by

$$f_k(\zeta) = \frac{\overline{\alpha_k}}{|\alpha_k|}\frac{\zeta - \alpha_k}{1 - \overline{\alpha_k}\zeta} + 1, \quad \zeta \in \Delta.$$

A straightforward computation using

$$\frac{1}{\lambda_{\Delta_1}(w)} = 2\,\mathrm{Re}\,w - |w|^2, \quad w \in \Delta_1,$$

yields

$$\lim_{k \to \infty} L_n(f_k, z, \Delta, \Delta_1) = \lim_{k \to \infty}\frac{(1 - |z|^2)^{n-1}}{|1 - \overline{\alpha_k}z|^{n-1}} = (1 + |z|)^{n-1}.$$

Theorem 3.5 assures that

$$1 - \frac{|w|^2}{2\,\mathrm{Re}\,w} = \frac{\lambda_{H_1}(w)}{\lambda_{\Delta_1}(w)} \leq \frac{\lambda_{\Pi}(w)}{\lambda_{\Delta_1}(w)} \leq 1, \quad w \in \Delta_1.$$

Since

$$\lim_{k \to \infty}\frac{|f_k(z)|^2}{2\,\mathrm{Re}\,(f_k(z))} = 0,$$

it is evident that

$$\lim_{k \to \infty} L_n(f_k, z, \Delta, \Pi) = \lim_{k \to \infty} L_n(f_k, z, \Delta, \Delta_1)\frac{\lambda_{\Pi}(f_k(z))}{\lambda_{\Delta_1}(f_k(z))} = (1 + |z|)^{n-1}.$$

This completes the proof of Theorem 4.13. $\qquad\Box$

Finally, we consider pairs (Ω, Π) with simply connected Ω and convex Π. We will need the constant

$$C_n(H_2, \Lambda) = \binom{2n-1}{n}$$

from Theorem 4.9 of Yamashita. By mathematical induction and Stirling's formula it can be shown that

$$\binom{2n-1}{n} = 4^{n-1/2}\frac{(2n-1)!!}{(2n)!!} = \frac{4^{n-1/2}}{\sqrt{n}}\left(\frac{1}{\sqrt{\pi}} + O(1/n)\right) \quad \text{as } n \to \infty,$$

and that

$$2^{n-1} < 4^{n-1/2}\frac{(2n-1)!!}{(2n)!!} < \frac{4^{n-1/2}}{\sqrt{n+1}} \quad (n \geq 2).$$

Theorem 4.9 is a special case of the following assertion.

Theorem 4.14 ([17]). *Let Ω and Π be simply connected proper subdomains of $\mathbb{C}$. If Π is a convex domain, then*

$$C_n(\Omega, \Pi) \leq C_n(H_2, \Pi) = 4^{n-1/2} \frac{(2n-1)!!}{(2n)!!}.$$

Proof. We follow the proof of our Theorem 4.10 with the same notation but some little changes.

Since Π is a convex domain and $u \prec t$, we can use Theorem 4.12 of Rogosinski to obtain that $|a_k| \lambda_\Pi(f(z_0)) \leq 1$. This together with formula (4.5) give

$$\frac{|f^{(n)}(z_0)|}{n!} \frac{\lambda_\Pi(f(z_0))}{(\lambda_\Omega(z_0))^n} \leq \sum_{k=1}^{n} |A_{n,k}(z_0)|.$$

As in the proof of Theorem 4.10 we get

$$\frac{|f^{(n)}(z_0)|}{n!} \frac{\lambda_\Pi(f(z_0))}{(\lambda_\Omega(z_0))^n} \leq \sum_{k=1}^{n} A_{n,k}(K_1).$$

Summing up gives

$$\sum_{k=1}^{n} A_{n,k}(K_1) = \frac{1}{2\pi i} \int_{\partial \Delta_r} \frac{(1+\zeta)^{2n-1}(1-\zeta)^n}{\zeta^n} \, d\zeta = \binom{2n-1}{n},$$

which proves the desired inequality

$$C_n(\Omega, \Pi) \leq \binom{2n-1}{n} = 4^{n-1/2} \frac{(2n-1)!!}{(2n)!!}.$$

For $\Pi = H_1$ the lower estimate

$$C_n(H_2, \Pi) \geq 4^{n-1/2} \frac{(2n-1)!!}{(2n)!!}$$

is given by the example of $f(\zeta) = \zeta^{-1/2}$ at the points $\zeta = z = x > 0$. To obtain the lower estimate for any convex domain Π, we proceed by approximation as in the proof of Theorem 4.13. Namely, we consider

$$\Delta_1 = \{w \mid |w - 1| < 1\} \subset \Pi \subset H_1$$

and the function $f \in A(H_2, \Delta_1) \subset A(H_2, H_1)$ defined by

$$f(\zeta) = \frac{2}{\sqrt{\zeta} + 1}, \quad \sqrt{1} = 1.$$

Straightforward computations using the comparison theorem and the asymptotic behavior of the nth derivative of this function at the point $z = x > 0$, as $x \to +\infty$,

$$\frac{x^{n+1/2} |f^{(n)}(x)|}{n!} = 2 \frac{(2n-1)!!}{(2n)!!} + o(1)$$

gives that

$$\lim_{x \to +\infty} \frac{|f^{(n)}(x)| \lambda_\Pi(f(x))}{n! \, (\lambda_{H_2}(x))^n} = 4^{n-1/2} \frac{(2n-1)!!}{(2n)!!}.$$

This completes the proof of Theorem 4.14. $\qquad\qquad\square$

4.5 Punishing factors for convex pairs

The main aim of the present section is to prove that $C_n(\Omega, \Pi) = 2^{n-1}$ for any pair (Ω, Π) of convex domains (see [24], see also Chua [56], Li [102], [103], and Yamashita [169] for special cases).

Theorem 4.15 ([24]). *Let Ω and Π be two convex proper subdomains of $\mathbb{C}$ and let $f \in A(\Omega, \Pi), n \in \mathbb{N}$. Then for any $z_0 \in \Omega$ the inequality*

$$\frac{|f^{(n)}(z_0)|}{n!} \le 2^{n-1} \frac{(\lambda_\Omega(z_0))^n}{\lambda_\Pi(f(z_0))} \tag{4.10}$$

is valid. The constant 2^{n-1} can not be replaced by a smaller one independent of $f \in A(\Omega, \Pi)$ and $z_0 \in \Omega$ for any pair (Ω, Π) of convex domains.

Proof. In the following we consider only the cases $n \ge 2$, since the case $n = 1$ is given by the Schwarz-Pick lemma. We first prove that $C_n(\Omega, \Pi) \le 2^{n-1}$. According to the central formula (4.5) we have to prove that

$$\left| \sum_{k=1}^n c_k A_{n,k} \right| \le 2^{n-1}.$$

In the present theorem, the functions s and t belong to the class K of functions univalent in Δ that map Δ onto a convex domain and are normalized as usual, e.g., $t(0) = 0$ and $t'(0) = 1$. Moreover, c_k are the Taylor coefficients of a holomorphic function subordinate to the function t, and $A_{n,k}$ is the n-th Taylor coefficient of the power $F^k(w)$, where F is the inverse of the function s. Since the extreme points of the closed convex hull of the set

$$\{u \mid u \prec t \text{ for some } t \in K\}$$

are the functions

$$\frac{xz}{1 - yz}, \quad z \in \Delta,$$

for fixed $(x, y) \in \partial\Delta \times \partial\Delta$ (see [75], Theorem 5.21), it remains to prove that

$$\mathrm{Re}\left(\sum_{k=1}^n xy^k A_{n,k} \right) \le \left| \sum_{k=1}^n y^k A_{n,k} \right| \le 2^{n-1}, \quad y \in \partial\Delta.$$

Let $a_{n-k,n}$ be defined by the Taylor expansion

$$\left(\frac{z}{s(z)}\right)^n = \sum_{\nu=0}^{\infty} a_{\nu,n} z^\nu.$$

Then the Schur-Jabotinsky theorem (compare for example [81], Theorem 1.9.a) implies that for $1 \le k \le n$ the identities

$$A_{n,k} = \frac{k}{n}\, a_{n-k,n}$$

are valid. Hence, we have to prove that

$$\left|\sum_{l=0}^{n-1} \frac{n-l}{n}\, y^{n-l}\, a_{l,n}\right| \le 2^{n-1}, \quad y \in \partial\Delta. \tag{4.11}$$

The tool for the proof of (4.11) is the resulting representation

$$\left(\frac{z}{s(z)}\right)^n = (1+z\omega(z))^n = 1 + \sum_{\sigma=1}^{n} \binom{n}{\sigma} z^\sigma (\omega(z))^\sigma, \quad z \in \Delta,$$

where $|\omega(z)| \le 1$ for any $z \in \Delta$ by the Marx-Strohhäcker theorem, that indicates $\mathrm{Re}(s(z)/z) > 1/2$ in Δ (see [115] and [155], see also Section 7.3 below). If we define

$$(\omega(z))^\sigma = \sum_{j=0}^{\infty} d_{j,\sigma} z^j, \quad z \in \Delta,$$

we get the following formula for the sum appearing in (4.11):

$$\sum_{l=0}^{n-1} \frac{n-l}{n}\, y^{n-l}\, a_{l,n} = y^n + \sum_{\sigma=1}^{n-1} \frac{1}{n}\binom{n}{\sigma}\sum_{j=\sigma}^{n-1}(n-j)y^{n-j}d_{j-\sigma,\sigma}. \tag{4.12}$$

To have the desired inequality it is sufficient to prove that

$$\left|\sum_{j=\sigma}^{n-1}(n-j)y^{n-j}d_{j-\sigma,\sigma}\right| \le n-\sigma.$$

Now we observe that the functions ω^σ also map the disc Δ into $\overline{\Delta}$. Therefore, we may replace the coefficients $d_{j-\sigma,\sigma}$ by the coefficients $d_{j-\sigma}$ of a unimodular bounded function when we consider the latter estimate. Taking $p = n - \sigma$ gives us an equivalent inequality

$$\left|\sum_{\tau=0}^{p-1}(p-\tau)\, y^{p-\tau} d_\tau\right| \le p.$$

This inequality follows directly from Fejér's inequality

$$\left| \sum_{\tau=0}^{p-1} (p-\tau) d_\tau \right| \le p,$$

which has long been known to be valid (see [61] and [159]). This concludes the proof of the inequality (4.11).

Now, we shall prove that the constant 2^{n-1} is best possible in any of the cases in question.

Lemma 4.16. *Let Ω and Π be two convex proper subdomains of $\mathbb{C}$. Then for any $n \ge 2$ the inequality*

$$C_n(\Omega, \Pi) \ge 2^{n-1}$$

is valid.

Proof. We know that the constant $C_n(\Omega, \Pi)$ is invariant under linear transformations of Ω and Π. Hence, without restriction of generality, we may assume that

$$\Delta_1 = \{z \mid |z-1| < 1\} \subset \Omega \subset H_1 = \{z \mid \operatorname{Re} z > 0\} \tag{4.13}$$

and

$$\Delta_1 \subset \Pi \subset H_1. \tag{4.14}$$

Let $\alpha \in (0,1)$ and $\xi \in (0,1)$ and consider the function

$$f_\alpha(z) = \alpha \, \frac{z+2}{z+\alpha}, \quad z \in \Omega.$$

Obviously, $f_\alpha \in A(\Omega, \Pi)$. By the comparison theorem applied to the inclusion relations (4.13) and (4.14) we get

$$\lim_{\beta \to 0+} \lambda_\Omega(\beta) \, 2\beta = \lim_{\beta \to 0+} \lambda_\Pi(\beta) \, 2\beta = 1.$$

Now, by use of these asymptotic equalities, we prove Lemma 4.16 with the following chain of inequalities and equations.

$$
\begin{aligned}
C_n(\Omega, \Pi) &\ge \lim_{\xi \to 0+} \lim_{\alpha \to 0+} \frac{\left| f_\alpha^{(n)}(\xi) \right|}{n!} \frac{\lambda_\Pi\left(f_\alpha(\xi)\right)}{\left(\lambda_\Omega(\xi)\right)^n} \\
&= \lim_{\xi \to 0+} \lim_{\alpha \to 0+} \frac{\left| f_\alpha^{(n)}(\xi) \right|}{n!} \frac{(2\xi)^n}{2\alpha \frac{\xi+2}{\xi+\alpha}} \\
&= \lim_{\xi \to 0+} \lim_{\alpha \to 0+} \frac{\alpha(2-\alpha)}{(\xi+\alpha)^{n+1}} \frac{(2\xi)^n}{2\alpha \frac{\xi+2}{\xi+\alpha}} = \lim_{\xi \to 0+} \frac{2^n}{2+\xi} = 2^{n-1}.
\end{aligned}
$$

$\square$

This concludes the proof of Theorem 4.15. $\square$

Remark 4.17. The last part of the proof shows that the constant 2^{n-1} is approached for any pair of convex domains, when z_0 and $f(z_0)$ approach the boundaries of Ω and Π at certain points. But there are simple special cases where the constant is attained at inner points. For instance, this happens if Ω and Π are half planes. Actually, if $\Omega = \Pi = H_1$ and $f_0(z) = 1/z$, then, at any point $z_0 = x > 0$,

$$\frac{f_0^{(n)}(x)}{n!} = \frac{1}{x^{n+1}} = \frac{2^{n-1}(\lambda_\Omega(x))^n}{\lambda_\Pi(1/x)},$$

since $1/\lambda_{H_1}(z) = 2\,\mathrm{Re}\,z$.

4.6   Case $n = 2$ for all domains

Let Ω and Π be hyperbolic domains on the Riemann sphere $\overline{\mathbb{C}}$ that are equipped with the Poincaré metric of curvature -4. According to Poincaré's generalization of Riemann's mapping theorem, this means that the boundaries of Ω and Π contain at least three points in $\overline{\mathbb{C}}$ and that the density of this metric in the unit disc $\Delta = \{z \mid |z| < 1\}$ is defined as

$$\lambda_\Delta(z) = \frac{1}{1 - |z|^2}, \quad z \in \Delta.$$

We have considered the functionals L_n, M_n and C_n for several special cases. According to [16], in the general case these functionals depend on the hyperbolic characteristics

$$(\lambda_\Omega(z))^{-k} \frac{\partial^k \log(\lambda_\Omega(z))}{\partial \overline{z}^k} \quad \text{and} \quad (\lambda_\Pi(w))^{-k} \frac{\partial^k \log(\lambda_\Pi(w))}{\partial \overline{w}^k}, \quad 1 \le k \le n-1.$$

Everything is clear in the case $n = 2$. In particular, the following theorem describes $C_2(\Omega, \Pi)$ in terms of the gradients

$$\nabla(1/\lambda_\Omega(z)) = -2(\lambda_\Omega(z))^{-1} \frac{\partial \log(\lambda_\Omega(z))}{\partial \overline{z}}$$

and

$$\nabla(1/\lambda_\Pi(w)) = -2(\lambda_\Pi(w))^{-1} \frac{\partial \log(\lambda_\Pi(w))}{\partial \overline{w}}.$$

Theorem 4.18 ([27]). *For all hyperbolic domains $\Omega \subset \overline{\mathbb{C}}$ and $\Pi \subset \overline{\mathbb{C}}$,*

$$C_2(\Omega, \Pi) = C_2(\Pi, \Omega) = \frac{1}{2}\left(\sup_{z \in \Omega} |\nabla(1/\lambda_\Omega(z))| + \sup_{w \in \Pi} |\nabla(1/\lambda_\Pi(w))|\right).$$

Proof. Fix $(z, w) \in \Omega \times \Pi$, $z \ne \infty$ and $w \ne \infty$. We consider a function $f \in A(\Omega, \Pi)$ such that $f(z) = w$. Let $\Phi : \Delta \to \Omega$ and $\Psi : \Delta \to \Pi$ be universal covering maps such that $\Phi(0) = z$ and $\Psi(0) = w$. We consider the holomorphic function $g : \Delta \to \Delta$ defined by

$$g(\zeta) = \Psi^{-1}(f(\Phi(\zeta))) = \sum_{n=1}^{\infty} c_n \zeta^n, \quad \zeta \in \Delta.$$

By direct computations, one gets the identity

$$\frac{f''(z)}{2} \frac{(\Phi'(0))^2}{\Psi'(0)} = c_2 + \frac{\Psi''(0)}{2\Psi'(0)} c_1^2 - \frac{\Phi''(0)}{2\Phi'(0)} c_1,$$

and therefore

$$\frac{|f''(z)|}{2} \frac{\lambda_\Pi(w)}{(\lambda_\Omega(z))^2} = \left| c_2 + \frac{y}{2} c_1^2 + \frac{x}{2} c_1 \right|, \tag{4.15}$$

where

$$x = -\frac{\Phi''(0)}{\Phi'(0)} \quad \text{and} \quad y = \frac{\Psi''(0)}{\Psi'(0)}.$$

If $g \in A(\Delta, \Delta)$ and $g(0) = 0$, then the function $f = \Psi \circ g \circ \Phi^{-1}$ belongs to the set $A(\Omega, \Pi)$ and $f(z) = w$. Therefore, we have to find

$$S(x,y) = \max \left\{ \left| c_2 + \frac{y}{2} c_1^2 + \frac{x}{2} c_1 \right| \mid g \in A(\Delta, \Delta) \text{ with } g(0) = 0 \right\}. \tag{4.16}$$

Using classical results on the Taylor coefficients of unimodular bounded functions (see [148]) we get

$$S(x,y) = \max \left\{ \left| c_2 + \frac{y}{2} c_1^2 + \frac{x}{2} c_1 \right| \mid |c_1| \leq 1, |c_2| \leq 1 - |c_1|^2 \right\}$$

and, by a little analysis,

$$S(x,y)) = \max \left\{ 1 - t^2 + \frac{|y|}{2} t^2 + \frac{|x|}{2} t \mid t \in [0,1] \right\} = \frac{1}{2} F(|x|, |y|),$$

where the function F is defined as

$$F(p,q) := \begin{cases} p + q, & \text{if } p + 2q \geq 4, \\ 2 + \frac{p^2}{8 - 4q}, & \text{if } p + 2q < 4, \end{cases}$$

for $p \geq 0$ and $q \geq 0$. Combining this with (4.15) and (4.16) one gets

$$\max\{|f''(z)| \mid f \in A(\Omega, \Pi), \text{ fixed } w = f(z)\} = F(|x|, |y|) \frac{(\lambda_\Omega(z))^2}{\lambda_\Pi(w)}, \tag{4.17}$$

where

$$|x| = p = |\nabla(1/\lambda_\Omega(z))| \quad \text{and} \quad |y| = q = |\nabla(1/\lambda_\Pi(w))|.$$

For a domain Ω, by formula (3.6) in the form

$$|\nabla\left(1/\lambda_\Omega(z)\right)| = \left|(1 - |\zeta|^2)\frac{\Phi''(\zeta)}{\Phi'(\zeta)} - 2\overline{\zeta}\right|$$

and by Theorem 3.23 it is known that

$$\sup_{z\in\Omega}|\nabla\left(1/\lambda_\Omega(z)\right)| = 2$$

if and only if Ω is convex and that

$$\sup_{z\in\Omega}|\nabla\left(1/\lambda_\Omega(z)\right)| > 2$$

in all other cases. Accordingly, for all hyperbolic domains Ω and Π,

$$\sup_{z\in\Omega}|\nabla\left(1/\lambda_\Omega(z)\right)| \geq 2, \quad \sup_{w\in\Pi}|\nabla\left(1/\lambda_\Pi(w)\right)| \geq 2.$$

Moreover, we remark that, for $a \geq 2$ and $b \geq 2$, one easily gets the identity

$$\max\{F(p,q) \mid p \in [0,a], \text{ and } q \in [0,b]\} = a + b,$$

since the condition $p + 2q < 4$ implies the chain of inequalities

$$2 + \frac{p^2}{8 - 4q} < 2 + \frac{p}{2} < a + b.$$

These together with formula (4.17) imply the assertion of Theorem 4.18. $\square$

Using Theorem 3.23 and Theorem 4.18 we obtain

Corollary 4.19. *Suppose that Ω and Π are hyperbolic domains in $\mathbb{C}$. Then the equation $C_2(\Omega,\Pi) = 2$ is valid if and only if Ω and Π are convex domains in $\mathbb{C}$.*

Using formulas (3.6) and (3.42) to compute the above gradients, one easily gets the following corollaries of Theorem 4.18.

Corollary 4.20. *Let $\alpha \in [1,2]$, and let $g(\zeta) = (\zeta+1)^\alpha$, $g(0) = 1$. If $\Omega_\alpha := g(\Delta)$, then $C_2(\Omega_\alpha, \Omega_\alpha) = 2\alpha$.*

Observe that (Ω_2, Ω_2) is an extremal pair in Theorem 4.10 and compare Section 5.5.

Corollary 4.21. *Let A_1 and A_2 be annuli with moduli M_1 and M_2, respectively. Then*

$$C_2(A_1, A_2) = \sqrt{1 + 4M_1^2} + \sqrt{1 + 4M_2^2}.$$

Using Theorem 3.12, equation (3.37) and Theorem 4.18 we immediately obtain the following assertion.

Corollary 4.22. *Suppose that Ω and Π are hyperbolic domains in $\mathbb{C}$. The constant $C_2(\Omega,\Pi)$ is finite if and only if Ω and Π both are domains with uniformly perfect boundary.*

If $\infty \in \Omega$ or $\infty \in \Pi$ and $n \geq 2$, then it is easy to verify that $C_n(\Omega,\Pi) = \infty$.

Chapter 5

Punishing factors for special cases

After a colloquium talk of the second author on estimates of the form

$$\frac{|f^{(n)}(z)|}{n!} \leq C_n(\Omega, \Pi)\frac{(\lambda_\Omega(z))^n}{\lambda_\Pi(f(z))}, \quad f \in A(\Omega, \Pi), \ z \in \Omega, \tag{5.1}$$

for simply connected domains Ω and Π in $\mathbb{C}$, Ch. Pommerenke ([132]) proposed to look at (5.1) in the following way. The quotient $(\lambda_\Omega(z))^n/\lambda_\Pi(f(z))$ reflects the influence of the positions of the points z and $f(z)$ in Ω and Π on the nth derivative $f^{(n)}(z)$, whereas the quantities $C_n(\Omega, \Pi)$ are factors punishing bad behaviour of Ω or Π at the boundary. This motivates the title of the present chapter as well as the titles of some our papers.

5.1 Solution of the Chua conjecture

In [56] Chua published the following conjecture among others.

Chua's conjecture. *Let Ω be a convex proper subdomain of $\mathbb{C}$ and let f be holomorphic and injective on Ω. Then for any $z \in \Omega$ and any $n \geq 2$ the inequality*

$$\left|\frac{f^{(n)}(z)}{f'(z)\,n!}\right| \leq (n+1)\,2^{n-2}\,(\lambda_\Omega(z))^{n-1} \tag{5.2}$$

holds true.

In [56], Chua settled this conjecture for $n = 2, 3, 4$ (see [102] and [103] for the cases $n = 5, 6, 7, 8$). Also, taking the limit $z \to 1$, $z \in (0, 1)$, in (5.2) for the Koebe function shows that the constant $(n + 1)2^{n-2}$ on the right side of (5.2) can not be replaced by a smaller one.

In the paper [24] we proved that for Ω convex, Π linearly accessible, and $n \geq 2$, the inequality

$$C_n(\Omega, \Pi) \leq (n+1)2^{n-2} \qquad (5.3)$$

is valid. The equation

$$|f'(z)| = \frac{\lambda_\Omega(z)}{\lambda_\Pi(f(z))}, \quad z \in \Omega,$$

holds for functions f injective on Ω. Hence, the inequality (5.3) implies the validity of Chua's conjecture for f that map Ω conformally onto a linearly accessible domain Π. But, in fact more is true.

Theorem 5.1 ([25]). *Let Ω be a convex proper subdomain of $\mathbb{C}$, and Π be a simply connected proper subdomain of $\mathbb{C}$. Let further $n \geq 2$. Then the inequality (5.3) is valid.*

Proof. Because of the central formula (4.5), for $n \geq 2$ it is evident that (5.3) will follow from the inequality

$$\left| \sum_{k=1}^{n} c_k A_{n,k} \right| \leq (n+1)2^{n-2}, \qquad (5.4)$$

where $A_{n,k}$ are the coefficients connected with the inverse of a function $s \in K$, and c_k are described by the condition that the sum of the series

$$\sum_{k=1}^{\infty} c_k z^k =: g_1(z), \quad z \in \Delta, \qquad (5.5)$$

is subordinate to a member of the family S.

Because of the Schur-Jabotinsky theorem and the Marx-Strohhäcker inequality we have to prove that

$$\left| \sum_{l=0}^{n-1} \frac{n-l}{n} c_{n-l} \, a_{l,n} \right|$$

$$= \left| c_n + \sum_{\sigma=1}^{n-1} \frac{1}{n} \binom{n}{\sigma} \sum_{j=\sigma}^{n-1} (n-j) c_{n-j} d_{j-\sigma,\sigma} \right| \leq (n+1)2^{n-2}. \qquad (5.6)$$

We may replace the coefficients $d_{j-\sigma,\sigma}$ by the coefficients $d_{j-\sigma}$ of a unimodular bounded function when we estimate the modulus of the inner sum in (5.6). Using

$$\sum_{\sigma=1}^{n} \binom{n}{\sigma} \frac{\sigma^2}{n} = (n+1)2^{n-2},$$

it is easily seen that the desired inequality follows from the inequality

$$\left| \sum_{j=\sigma}^{n-1} (n-j) c_{n-j} d_{j-\sigma,\sigma} \right| \leq (n-\sigma)^2.$$

This is a consequence of the following lemma with $p = n - \sigma$. $\quad\square$

Lemma 5.2. *Let*

$$\tilde{\omega}(z) = \sum_{\tau=0}^{\infty} d_\tau z^\tau$$

be holomorphic in the unit disc and such that $\tilde{\omega}(\Delta) \subset \overline{\Delta}$ and let g_1 be the function defined by the equation (5.5) and subordinate to a function from S. Then for $p \in \mathbb{N}$ the inequality

$$\left| \sum_{\tau=0}^{p-1} (p-\tau)\, c_{p-\tau} d_\tau \right| \leq p^2 \tag{5.7}$$

is valid.

Proof. We shall use Sheil-Small's theorem 2.9 which says that

If g_1 is subordinated to a function $g \in S$ and P is a polynomial of degree $\leq p$, then for $z \in \overline{\Delta}$ the inequality

$$|(P * g_1)(z)| \leq p \max\{|P(z)|,\, |z| = 1\} \tag{5.8}$$

is valid.

To prove (5.7) with the help of (5.8) we consider the polynomial

$$P(z) = \sum_{\tau=0}^{p-1} d_\tau (p-\tau) z^{p-\tau}.$$

Because of the identity

$$(P * g_1)(1) = \sum_{\tau=0}^{p-1} (p-\tau)\, c_{p-\tau} d_\tau,$$

it is sufficient for the proof of (5.7) to show that

$$\left| P\left(e^{-i\theta}\right) \right| \leq p, \quad \theta \in [0, 2\pi].$$

Since the family of functions $\tilde{\omega}$ is invariant against rotations of the unit disc, it remains to prove the inequality

$$\left| \sum_{\tau=0}^{p-1} (p-\tau) d_\tau \right| \leq p.$$

This exactly is Fejér's inequality, used above. This concludes the proof of Lemma 5.2. $\quad\square$

Theorem 5.1 follows immediately.

5.2   Punishing factors for angles

We again consider the quantities $L_n(f, z, \Omega, \Pi)$, $M_n(z, \Omega, \Pi)$ and $C_n(\Omega, \Pi)$ defined by

$$\frac{1}{n!} \left| f^{(n)}(z) \right| = L_n(f, z, \Omega, \Pi) \frac{(\lambda_\Omega(z))^n}{\lambda_\Pi(f(z))}, n \in \mathbb{N}, f \in A(\Omega, \Pi), z \in \Omega,$$

$$M_n(z, \Omega, \Pi) := \sup\{L_n(f, z, \Omega, \Pi) \mid f \in A(\Omega, \Pi)\},$$

and

$$C_n(\Omega, \Pi) := \sup\{M_n(z, \Omega, \Pi) \mid z \in \Omega\}.$$

In this section we are concerned with the above quantities where one or both of the domains in question belong to the class of angular domains $aH_\alpha + b$ with opening angle $\alpha\pi$, $1 \le \alpha \le 2$, which means that there exists a linear transformation $T(z) = az + b$ such that $aH_\alpha + b := T(H_\alpha)$, where

$$H_\alpha = \left\{ z \mid |\arg z| < \frac{\alpha\pi}{2} \right\}.$$

Theorem 5.3 ([17]). *Let $1 \le \alpha \le 2$, $n \in \mathbb{N}$, and $z_0 \in \Delta$. Then*

$$M_n(z_0, \Delta, H_\alpha) = \sum_{k=0}^{n-1} (1 + |z_0|)^k \left(\begin{array}{c} n-1 \\ k \end{array} \right) \left(\begin{array}{c} \alpha \\ n-k \end{array} \right) \frac{2^{n-k-1}}{\alpha},$$

where

$$\left(\begin{array}{c} \alpha \\ \nu \end{array} \right) = \frac{1}{\nu!} \prod_{\mu=0}^{\nu-1} (\alpha - \mu).$$

If we let $|z_0| \to 1$, then we immediately obtain the following result.

Corollary 5.4. *Let $1 \le \alpha \le 2$ and $n \in \mathbb{N}$. Then*

$$C_n(\Delta, H_\alpha) = \frac{2^{n-1}}{\alpha} \left(\begin{array}{c} n+\alpha-1 \\ n \end{array} \right).$$

Proof of Theorem 5.3. The formula (4.3) indicates that we may insert into (4.5) for $\Omega = \Delta$ the identities

$$A_{n,k}(z_0) = \left(\begin{array}{c} n-1 \\ n-k \end{array} \right) \overline{z_0}^{n-k}.$$

Next, we set in Theorem 2.19,

$$g(\zeta) = f \left(\frac{\zeta + z_0}{1 + \overline{z_0}\zeta} \right),$$

and define the Taylor coefficients $d_k(\alpha)$ by

$$\frac{1}{2\alpha}\left(\left(\frac{1+\zeta}{1-\zeta}\right)^\alpha - 1\right) = \sum_{k=1}^\infty d_k(\alpha)\zeta^k.$$

Since the coefficients $d_k(\alpha)$ are nonnegative in our cases, we see that in (4.5) we may use the inequalities

$$|a_k|\lambda_{H_\alpha}(f(z_0)) \le d_k(\alpha)$$

to get

$$M_n(z_0, \Delta, H_\alpha) \le \sum_{k=1}^n \binom{n-1}{n-k} |z_0|^{n-k} d_k(\alpha). \tag{5.9}$$

It is easily seen that in this inequality the upper bound is attained for $z_0 = r_0 e^{i\theta}$, $r_0 \ge 0$, if we choose f to be given by the identity

$$\left(f\left(\frac{\zeta + z_0}{1 + \bar{z}_0\zeta}\right) - f(z_0)\right)\lambda_{H_\alpha}(f(z_0)) = \frac{e^{i\theta}}{2\alpha}\left(\left(\frac{1 + e^{-i\theta}\zeta}{1 - e^{-i\theta}\zeta}\right)^\alpha - 1\right).$$

It remains to show that the upper bound for $M_n(z_0, \Delta, H_\alpha)$ in (5.9) equals the upper bound given in Theorem 5.3. To this end we remark that the left side of the inequality (5.9) is the nth Taylor coefficient of the function

$$\frac{(1 + |z_0|z)^{n-1}}{2\alpha}\left(\frac{1+z}{1-z}\right)^\alpha.$$

Use of the binomial theorem for

$$(1 + |z_0|z)^{n-1} = ((1 + |z_0|)z + 1 - z)^{n-1}$$

reveals that the said nth Taylor coefficient equals the nth Taylor coefficient of the product

$$\frac{1}{2\alpha}(1+z)^\alpha \sum_{k=1}^n \binom{n-1}{n-k} z^{k-1}(1 + |z_0|)^{k-1}(1-z)^{n-k-\alpha}.$$

Hence, we have to sum up the $(n - k + 1)$th Taylor coefficients of the functions

$$\frac{1}{2\alpha}(1+z)^\alpha \binom{n-1}{n-k}(1 + |z_0|)^{k-1}(1-z)^{n-k-\alpha}, \quad k = 1, \ldots, n.$$

These coefficients may be written in the form

$$\binom{n-1}{n-k}\sum_{q=0}^{n-k+1}\binom{\alpha}{q}\frac{(-1)^{n-k+1-q}}{2\alpha}\binom{n-k-\alpha}{n-k+1-q}(1 + |z_0|)^{k-1}$$

$$= (1 + |z_0|)^{k-1}\binom{n-1}{n-k}\binom{\alpha}{n-k+1}\frac{2^{n-k}}{\alpha}, \quad k = 1, \ldots, n.$$

This proves the result of Theorem 5.3 if we replace k by $k - 1$ and sum up from $k = 0$ to $k = n - 1$. $\qquad\square$

Using theorems on inverse coefficients of conformal mappings we generalize Corollary 5.4 and Yamashita's results in [170] (see Theorem 4.9) as follows.

Theorem 5.5 ([17]).   a) *Let* $1 \leq \alpha, \beta \leq 2$ *and* $n \in \mathbb{N}$. *Then*

$$C_n(H_\beta, H_\alpha) = \frac{(2\beta)^n}{2\alpha} \binom{\frac{\alpha}{\beta} + n - 1}{n}.$$

b) *Let* $1 \leq \beta \leq 2$, $n \in \mathbb{N}$, *and* Π *be a convex proper subdomain of* $\mathbb{C}$. *Then*

$$C_n(H_\beta, \Pi) = C_n(H_\beta, H_1).$$

c) *Let* $1 \leq \beta \leq 2$, $n \in \mathbb{N}$, *and* Π *be a simply connected proper subdomain of* $\mathbb{C}$. *Then*

$$C_n(H_\beta, \Pi) \leq C_n(H_\beta, H_2).$$

d) *Let* $1 \leq \alpha \leq 2$, $n \in \mathbb{N}$, *and* Ω *be a simply connected proper subdomain of* $\mathbb{C}$. *Then*

$$C_n(\Omega, H_\alpha) \leq C_n(H_2, H_\alpha).$$

Remark 5.6. For all $\alpha, \beta \in [1, 2]$, one has that $C_1(H_\beta, H_\alpha) = 1$ in accordance with the Schwarz-Pick inequality. In the case $n \geq 2$,

$$C_n(H_\beta, H_\alpha) = \frac{2^n}{n!} \prod_{k=1}^{n-1} (\alpha + k\beta).$$

We see that $C_2(H_\beta, H_\alpha) = C_2(H_\alpha, H_\beta) = \alpha + \beta$ in accordance with Theorem 4.18, and

$$C_n(H_\alpha, H_\beta) > C_n(H_\beta, H_\alpha) > 2^{n-1},$$

whenever $n \geq 3$ and $\alpha > \beta \geq 1$.

Proof of Theorem 5.5. a) Firstly, we notice that the nth Taylor coefficient of $(h_\beta^{-1}(w))^k$ is nonnegative, since the Taylor coefficients of $h_\beta^{-1}(w)$ are all nonnegative. This fact and Theorem 2.20 imply that we now may use as an upper bound for the quantity $|A_{n,k}(z_0)|, 1 \leq k \leq n$, in formula (4.5) this nth Taylor coefficient of $(h_\beta^{-1}(w))^k$. For $|a_k|$ we use the same inequality as in the proof of Theorem 5.3. To prove that the resulting inequality is sharp and to compute the explicit formula given in the assertion, we use the conformal map $f_{\alpha,\beta}$ defined by

$$f_{\alpha,\beta}(z) := g_\alpha \left(h_\beta^{-1}(z) \right), \tag{5.10}$$

where

$$g_\alpha(\zeta) := \frac{1}{2\alpha} \left(\left(\frac{1+\zeta}{1-\zeta} \right)^\alpha - 1 \right), \quad \zeta \in \Delta.$$

One one hand, $f_{\alpha,\beta}$ maps the special angular domain

$$H_\beta = \left\{ z \,\middle|\, \left|\arg\left(z - \frac{1}{2\beta}\right)\right| < \frac{\beta\pi}{2} \right\}$$

conformally onto the special angular domain

$$H_\alpha = \left\{ z \,\middle|\, \left|\arg\left(z + \frac{1}{2\alpha}\right)\right| < \frac{\alpha\pi}{2} \right\}$$

such that $f_{\alpha,\beta}(0) = 0$. Now, we compute $f_{\alpha,\beta}^{(n)}(0)$ starting with formula (5.10) in the same way as we obtained (4.5). If we use that in our circumstances

$$\lambda_{H_\beta}(0) = \lambda_{H_\alpha}(0) = 1,$$

we observe that the resulting sum is just the sum that we derived as an upper bound in the beginning of this proof.

On the other hand, a direct computation of $f_{\alpha,\beta}$ shows that

$$f_{\alpha,\beta}(z) = \frac{1}{2\alpha}\left((1 - 2\beta z)^{-\frac{\alpha}{\beta}} - 1 \right).$$

Hence,

$$\frac{f_{\alpha,\beta}^{(n)}(0)}{n!} = \frac{(2\beta)^n}{2\alpha} \left(\begin{array}{c} \frac{\alpha}{\beta} + n - 1 \\ n \end{array} \right).$$

This proves the assertion in question.

b) The inequality

$$C_n(H_\beta, \Pi) \leq C_n(H_\beta, H_1)$$

for a convex domain Π follows in our case from Rogosinski's theorem 4.12 on convex subordination according to which we may apply

$$|a_k|\lambda_\Pi(z_0) \leq 1$$

in (4.5) in a way analogous to that of the proof of the case a).

The proof of

$$C_n(H_\beta, \Pi) \geq C_n(H_\beta, H_1)$$

for a convex domain follows the proof of

$$M_n(z, \Delta, \Pi) = (1 + |z|)^{n-1}$$

for convex Π in Chapter 4. Therefore, we only describe the crucial steps here. Without loss of generality, we may assume that

$$\Delta_1 = \{w \mid |w - 1| < 1\} \subset \Pi \subset H_1 = \{w \mid \operatorname{Re} w > 0\}.$$

For the special angular domain H_β chosen as in the proof of the case a) we consider the holomorphic functions

$$f_k(z) = \frac{2}{1 + k(1 - 2\beta z)^{\frac{1}{\beta}}}, \quad k \in \mathbb{N},$$

which map H_β conformally onto Δ_1 such that $f_k(0) = 2/(k+1) =: w_k$. We compute

$$\lambda_{H_\beta}(0) = 1, \quad \lambda_{\Delta_1}(w_k) = \frac{(k+1)^2}{4k}.$$

Since the sequences $(kf_k)_{k\in\mathbb{N}}$ are uniformly convergent on a neighbourhood of the origin we see that

$$\lim_{k\to\infty} k f_k^{(n)}(0) = 2 \left.\frac{d^n}{dz^n}(1 - 2\beta z)^{-\frac{1}{\beta}}\right|_{z=0} = 4\,(n!)C_n(H_\beta, H_1).$$

The last equation follows from the proof of the case a). Using the comparison principle for densities of the Poincarè metric (see Theoren 3.5) as in the proof of the corresponding relation in Chapter 4, it is easily seen that

$$\lim_{k\to\infty} \frac{\lambda_\Pi(w_k)}{\lambda_{\Delta_1}(w_k)} = 1.$$

Gathering together the above relations we get the equation

$$\lim_{k\to\infty} L_n(f_k, 0, H_\beta, \Pi) = \lim_{k\to\infty} \left(\frac{f^{(n)}(0)\lambda_{\Delta_1}(w_k)}{n!\left(\lambda_{H_\beta}(0)\right)^n} \frac{\lambda_\Pi(w_k)}{\lambda_{\Delta_1}(w_k)} \right) = C_n(H_\beta, H_1),$$

which immediately implies the assertion.

c) In this proof we only have to use again the validity of the Rogosinski or generalized Bieberbach conjecture together with the case a) of Theorem 5.5.

d) Here we use in the application of (4.5) that, according to Löwner's Theorem 2.13, the Taylor coefficients of the inverses of schlicht functions and their kth powers, $k \in \mathbb{N}$, are dominated by the related (positive) coefficients of the inverse of the Koebe function $z/(1+z)^2$ and its kth powers. This results in the assertion. □

In our next result we are concerned with proper subdomains Ω of $\overline{\mathbb{C}}$ simply connected with respect to $\overline{\mathbb{C}}$ and holomorphic functions $f : \Omega \to H_\alpha$.

In the sequel we use the quantity $p \in (0, 1]$ such that $p := p(z_0) = \tanh D_\Omega(z_0, \infty)$. By definition, $p = 1$, if $\infty \notin \Omega$.

The following theorem forms a bridge between the results for Π convex and Π simply connected with respect to $\mathbb{C}$.

Theorem 5.7 ([17]). *Let Ω be as above, $z_0 \in \Omega \setminus \{\infty\}$, $1 \leq \alpha \leq 2$, and $n \geq 2$. Then for any holomorphic function $f : \Omega \to H_\alpha$, the inequality*

$$\frac{1}{n!}|f^{(n)}(z_0)| \leq \frac{(\lambda_\Omega(z_0))^n}{\lambda_{H_\alpha}(f(z_0))} \sum_{k=1}^{n} \frac{4^k}{2\alpha} \binom{n-1}{n-k} \binom{\frac{\alpha}{2}}{k} \left(p(z_0) + \frac{1}{p(z_0)} + 2 \right)^{n-k}$$

is valid. For each $n \geq 2$, each $\alpha \in [1,2]$, and each $p \in (0,1]$ there exist $\Omega, H_\alpha, z_0 \in \Omega \setminus \{\infty\}$, and f as above such that equality is attained in the above inequality.

Proof. We again start with the central equality (4.5). Let

$$p = p(z_0), \quad a = p + \frac{1}{p}, \quad \kappa_p(\zeta) = \frac{\zeta}{(1+p\zeta)(1+\frac{\zeta}{p})}, \quad \zeta \in \Delta,$$

and K_p the function inverse to κ_p having an expansion

$$K_p(z) = z + \sum_{\nu=2}^{\infty} B_\nu z^\nu$$

valid in some neighbourhood of the origin. Considering the class S_p of functions meromorphic and univalent in Δ which have a pole in a point b, $|b| = p$, and an expansion

$$m(\zeta) = \zeta + \sum_{\nu=2}^{\infty} A_\nu \zeta^\nu, \quad |\zeta| < p,$$

Baernstein and Schober showed in [33] (see Theorem 2.15) that

$$|A_\nu| \leq B_\nu, \quad \nu \geq 2.$$

This implies that the nth Taylor coefficients of the kth powers of $m \in S_p$ are dominated by the nth Taylor coefficients $B_{n,k}$ of the kth powers of the function K_p. By a procedure analogous to that, which led us to formula (4.5), we get in our case for the meromorphic functions f in question the inequality

$$\frac{|f^{(n)}(z_0)|}{n!} \leq \frac{(\lambda_\Omega(z_0))^n}{\lambda_{H_\alpha}(f(z_0))} \sum_{k=1}^{n} B_{n,k} d_k(\alpha). \tag{5.11}$$

To show that this inequality is sharp and that the right-hand side has the form given in the assertion, we proceed in principle as in the proof of the case a) of Theorem 5.5. To this end we consider in the present case the function

$$f_{\alpha,p}(z) := g_\alpha(K_p(z))$$

which maps the domain

$$\Omega_0 = \overline{\mathbb{C}} \setminus [(a+2)^{-1}, (a-2)^{-1}]$$

conformally onto the special angular domain H_α chosen as in the proof . We see that in this case $f_{\alpha,p}(0) = 0$ and $\lambda_{H_\alpha}(0) = \lambda_{\Omega_0}(0) = 1$ and we deduce the sharpness of (5.11) considering $f_{\alpha,p}^{(n)}(0)$ as above. To get the explicit form of the upper bound we consider

$$f_{\alpha,p}(z) = \frac{1}{2\alpha}\left(\left(\frac{1-(a-2)z}{1-(a+2)z}\right)^{\frac{\alpha}{2}} - 1\right) =: \sum_{n=1}^{\infty} C_n(\alpha,p)z^n.$$

This implies

$$\left(1 + \sum_{k=1}^{\infty} 4(a+2)^{k-1}z^k\right)^{\frac{\alpha}{2}} = 1 + 2\alpha\sum_{n=1}^{\infty} C_n(\alpha,p)z^n.$$

With the abbreviation $Z := z(a+2)$ this is equivalent to the equation

$$1 + \sum_{k=1}^{\infty}\binom{\frac{\alpha}{2}}{k}\left(\frac{4}{a+2}\right)^k\left(\frac{Z}{1-Z}\right)^k = 1 + 2\alpha\sum_{n=1}^{\infty}\frac{C_n(\alpha,p)}{(a+2)^n}Z^n.$$

If we then expand $Z^k/(1-Z)^k$ in powers of Z and compare the coefficients on both sides of this equation, we then establish the assertion of Theorem 5.7 on the sharpness of the punishing factor. $\qquad\qquad\square$

5.3   Sharp lower bounds for punishing factors

We shall consider domains with a special local property at a boundary point, which will be a metrical characterization of the "bad" behavior of the boundary.

Let Ω be a domain in $\overline{\mathbb{C}}$. We denote

$$a\Omega + b := \{z = a\zeta + b \mid \zeta \in \Omega\}, \quad a, b \in \mathbb{C}, \ a \neq 0,$$

and

$$\frac{1}{\Omega} := \left\{z = \frac{1}{\zeta} \mid \zeta \in \Omega\right\}.$$

For $\alpha \in [1,2]$ we define

$$\Delta_\alpha^+ := \{z = (\zeta + 1/2)^\alpha \mid |\zeta| < 1/2\},$$

$$\Delta_0^- := [-1,0]$$

and

$$\Delta_{2-\alpha}^- := \left\{z = -\frac{w}{w+1} \mid w = \left(\frac{1-\zeta}{1+\zeta}\right)^{2-\alpha}, \ |\zeta| \leq 1\right\} \quad \text{for} \quad \alpha \in [1,2),$$

where the branch of a power is fixed by $1^\alpha = 1$. Observe that $z = 0$ is a boundary point for all sets Δ_α^+ and $\Delta_{2-\alpha}^-$.

Definition 5.8. Let Ω be a domain in $\overline{\mathbb{C}}$, let $\alpha \in [1,2]$, and let $z_0 \in (\partial\Omega) \setminus \{\infty\}$. If there exist $\varepsilon > 0$ and $t \in \mathbb{R}$ such that

$$\varepsilon e^{it} \Delta_\alpha^+ + z_0 \subset \Omega \quad \text{and} \quad \varepsilon e^{it} \Delta_{2-\alpha}^- + z_0 \subset \overline{\mathbb{C}} \setminus \Omega, \tag{5.12}$$

then we say that z_0 is an angular point on $\partial\Omega$ of order α.

Clearly, the condition (5.12) is equivalent to the inclusions

$$\Delta_\alpha^+ \subset a\Omega + b \subset \overline{\mathbb{C}} \setminus \Delta_{2-\alpha}^-, \qquad a = \frac{e^{-it}}{\varepsilon}, \quad b = -z_0 \frac{e^{-it}}{\varepsilon}.$$

We extend this definition to the boundary point at infinity.

Definition 5.9. Let $\alpha \in [1,2]$ and let Ω be a domain in $\mathbb{C}$ such that $\infty \in \partial\Omega$. If there exist complex numbers $a \neq 0$ and b such that

$$\Delta_\alpha^+ \subset \frac{1}{a\Omega + b} \subset \overline{\mathbb{C}} \setminus \Delta_{2-\alpha}^-,$$

then we say that the point at infinity is an angular point on $\partial\Omega$ of order α.

According to Definitions 5.8 and 5.9 the domain

$$H_\alpha := \left\{ z \in \mathbb{C} \,\middle|\, |\arg z| < \frac{\alpha\pi}{2} \right\}$$

has two boundary angular points of order α that are $z_0 = 0$ and the point at infinity. A non-trivial example is

$$\Omega_1 = H_1 \setminus \bigcup_{k=1}^{\infty} D_k,$$

where

$$D_k = \bigcup_{0 \le t < \infty} \{ z \in \mathbb{C} \mid |z - 1/k - it| \le (t + 2k)^{-2} \}.$$

It is easy to verify that

$$\Delta_1^+ \subset \frac{1}{\Omega_1 - 1} \subset \overline{\mathbb{C}} \setminus \Delta_1^-.$$

Clearly, $z_0 = \infty$ is an angular point on $\partial\Omega_1$ of order 1.

Theorem 5.10. *Let Ω and Π be domains in $\mathbb{C}$. If there exist boundary points such that $z_0 \in \overline{\mathbb{C}}$ is an angular point on $\partial\Omega$ of order $\alpha \in [1,2]$ and $w_0 \in \overline{\mathbb{C}}$ is an angular point on $\partial\Pi$ of order $\beta \in [1,2]$, then for any $n \in \mathbb{N}$,*

$$C_n(\Omega, \Pi) \ge C_n(H_\alpha, H_\beta) = \frac{(2\alpha)^n}{2\beta} \binom{\frac{\beta}{\alpha} + n - 1}{n}. \tag{5.13}$$

Proof. First we consider the case when $z_0 \neq \infty$ and $w_0 \neq \infty$. Because of the invariance of the punishing factor under linear transformations of domains, we can suppose that

$$\Delta_\alpha^+ \subset \Omega \subset \overline{\mathbb{C}} \setminus \Delta_{2-\alpha}^-, \quad \Delta_\beta^+ \subset \Pi \subset \overline{\mathbb{C}} \setminus \Delta_{2-\beta}^-. \tag{5.14}$$

Using the conformal mappings $g_1 : \Delta \to \Delta_\alpha^+$ and $g_2 : H_1 \to \overline{\mathbb{C}} \setminus \Delta_{2-\alpha}^-$ explicitly given by functions

$$g_1(\zeta) = \left(\frac{1+\zeta}{2}\right)^\alpha, \quad g_2(\zeta) = \frac{\zeta^\alpha}{1 - \zeta^\alpha}$$

and the conformal invariance of the Poincaré metric, we easily obtain that

$$\lambda_{\Delta_\alpha^+}(x) = \frac{1}{2\alpha x(1 - x^{1/\alpha})}, \quad \lambda_{\overline{\mathbb{C}} \setminus \Delta_{2-\alpha}^-}(x) = \frac{1}{2\alpha x(1 + x)}$$

for any $x = \mathrm{Re} z \in (0,1)$ and any $\alpha \in [1,2]$. These equations together with (5.14) and the comparison theorem give

$$\lambda_\Omega(x) = \frac{1}{2\alpha x + o(x)}, \quad \lambda_\Pi(x) = \frac{1}{2\beta x + o(x)} \quad \text{as } x \to 0^+. \tag{5.15}$$

For any $c \in (0, \infty)$, the function f_c defined by

$$f_c(z) = \frac{c^\beta \varphi^\beta(z)}{(1 + c\varphi(z))^\beta}, \quad \varphi(z) := \left(1 + \frac{1}{z}\right)^{1/\alpha},$$

$$z \in \overline{\mathbb{C}} \setminus \Delta_{2-\alpha}^-, \quad \varphi(1) := 2^{1/\alpha},$$

is a conformal map of $\overline{\mathbb{C}} \setminus \Delta_{2-\alpha}^-$ onto Δ_α^+. According to inclusions (5.14) we have that $f_c \in A(\Omega, \Pi)$ for any positive value of c. By the Leibniz formula, one gets

$$\frac{f_c^{(n)}(x)}{c^\beta} = \sum_{k=0}^{n} \binom{n}{k} \frac{d^k \varphi^\beta(x)}{dx^k} \frac{d^{n-k}(1 + c\varphi(x))^{-\beta}}{dx^{n-k}}$$

$$= \frac{d^n \varphi^\beta(x)}{dx^n}(1 + O(c)) \quad \text{as } c \to 0^+.$$

Using this, the formulas from (5.15) and the definition of $C_n(\Omega, \Pi)$ we obtain

$$C_n(\Omega, \Pi)$$

$$\geq \lim_{x \to 0^+} \lim_{c \to 0^+} \frac{\lambda_\Pi(f_c(x))}{(\lambda_\Omega(x))^n} \frac{|f_c^{(n)}(x)|}{n!}$$

$$= \lim_{x \to 0^+} \lim_{c \to 0^+} \frac{(2\alpha x)^n c^\beta}{2\beta f_c(x) n!} \left| \frac{d^n \varphi^\beta(x)}{dx^n} \right| = \frac{(2\alpha)^n}{2\beta n!} \lim_{x \to 0^+} x^{n+\beta/\alpha} \left| \frac{d^n (x^{-\beta/\alpha}(x))}{dx^n} \right|$$

$$= \frac{(2\alpha)^n}{2\beta} \binom{\frac{\beta}{\alpha} + n - 1}{n} = C_n(H_\alpha, H_\beta).$$

This completes the proof of the desired inequality (5.13) in the case $z_0 \in \mathbb{C}$ and, $w_0 \in \mathbb{C}$.

Now, we consider the case $(z_0, w_0) = (\infty, \infty)$. Without loss of generality we may suppose that

$$\Delta_\alpha^+ \subset \frac{1}{\Omega} \subset \overline{\mathbb{C}} \setminus \Delta_{2-\alpha}^-, \quad \Delta_\beta^+ \subset \frac{1}{\Pi} \subset \overline{\mathbb{C}} \setminus \Delta_{2-\beta}^-$$

or, equivalently,

$$\frac{1}{\Delta_\alpha^+} \subset \Omega \subset \frac{1}{\overline{\mathbb{C}} \setminus \Delta_{2-\alpha}^-} = H_\alpha - 1, \quad \frac{1}{\Delta_\beta^+} \subset \Pi \subset H_\beta - 1.$$

As members of $A(\Omega, \Pi)$ we choose conformal mappings $g_c : H_\alpha - 1 \to 1/\Delta_\beta^+$ defined by

$$g_c(z) = \frac{1}{f_c(1/z)} = \frac{(1 + c(1+z)^{1/\alpha})^\beta}{c^\beta (1+z)^{\beta/\alpha}}, \quad c \in (0, +\infty).$$

For any domain D it is evident that $\lambda_{1/D}(1/z) = \lambda_D(z)|z|^2$. Using this at $z = x \in (1, +\infty)$, the known formulas from the first part of this proof, and the comparison theorem, one easily gets

$$\lambda_\Omega(x) = \frac{1}{2\alpha x + o(x)} \quad \text{as } x \to +\infty$$

and

$$\lambda_\Pi(g_c(x)) = \frac{1}{2\beta x^{\alpha/\beta} c^{-\beta} + o(c^{-\beta})} \quad \text{as } c \to 0^+.$$

Since

$$\frac{g_c^{(n)}(x)}{n!} = \frac{1}{c^\beta n!} \frac{d^n (1+x)^{-\beta/\alpha}}{dx^n}(1 + O(c)) = \frac{(-1)^n}{c^\beta} \binom{\frac{\beta}{\alpha} + n - 1}{n}(1 + O(c)),$$

we immediately obtain

$$C_n(\Omega, \Pi) \geq \lim_{x \to +\infty} \lim_{c \to 0^+} \frac{\lambda_\Pi(g_c(x))}{(\lambda_\Omega(x))^n} \frac{|g_c^{(n)}(x)|}{n!} = C_n(H_\alpha, H_\beta).$$

If $z_0 \in \mathbb{C}$, $w_0 = \infty$, we consider the pair of angular points $(z_0, w_0) = (0, \infty)$ and domains such that

$$\Delta_\alpha^+ \subset \Omega \subset \overline{\mathbb{C}} \setminus \Delta_{2-\alpha}^-, \quad \frac{1}{\Delta_\beta^+} \subset \Pi \subset H_\beta - 1.$$

To obtain the lower estimate for the punishing factor, we proceed as in the case $(z_0, w_0) = (0, 0)$, choosing as members of $A(\Omega, \Pi)$ the conformal mappings

$$f_c(.; 0, \infty) : \overline{\mathbb{C}} \setminus \Delta_{2-\alpha}^- \to \frac{1}{\Delta_\beta^+}$$

defined by equations

$$f_c(z; 0, \infty) = 1/f_c(-z-1) = c^{-\beta} \left(c + (1 + 1/z)^{-1/\alpha} \right)^{\beta}.$$

Using the properties

$$\lambda_\Omega(x) = \frac{1}{2\alpha x + o(x)}, \quad \text{as } x \to 0^+, \quad \lambda_\Pi(x) = \frac{1}{2\beta x + o(x)} \quad \text{as } x \to +\infty,$$

one easily obtains

$$C_n(\Omega, \Pi) \geq \lim_{x \to 0^+} \lim_{c \to 0^+} \frac{\lambda_\Pi(f_c(x; 0, \infty))}{(\lambda_\Omega(x))^n} \frac{|f_c^{(n)}(x; 0, \infty)|}{n!} = C_n(H_\alpha, H_\beta).$$

If $z_0 = \infty$, $w_0 \in \mathbb{C}$, then the proof is similar. We consider the pair $(z_0, w_0) = (\infty, 0)$, domains with the property

$$\Delta_\beta^+ \subset \Pi \subset \overline{\mathbb{C}} \setminus \Delta_{2-\beta}^-, \quad \frac{1}{\Delta_\alpha^+} \subset \Omega \subset H_\alpha - 1,$$

and the functions defined by

$$f_c(z; \infty, 0) = f_c(-z-1) = \frac{c^\beta (1 + 1/z)^{-\beta/\alpha}}{\left(1 + c(1 + 1/z)^{-1/\alpha} \right)^\beta}.$$

Now, it is easy to verify that

$$C_n(\Omega, \Pi) \geq \lim_{x \to +\infty} \lim_{c \to 0^+} \frac{\lambda_\Pi(f_c(x; \infty, 0))}{(\lambda_\Omega(x))^n} \frac{|f_c^{(n)}(x; \infty, 0)|}{n!} = C_n(H_\alpha, H_\beta).$$

This completes the proof of Theorem 5.10.   □

Combining the case $\alpha = \beta = 2$ of this theorem with the known upper estimate for simply connected domains, one obtains

Corollary 5.11. *Let $n \in \mathbb{N}$ and let Ω and Π be simply connected domains in $\mathbb{C}$ such that there is an angular point of order 2 on each of the two boundaries $\partial\Omega$ and $\partial\Pi$. Then*

$$C_n(\Omega, \Pi) = 4^{n-1}.$$

Combining the case $\beta = 1$, $n = 2$ of Theorem 5.10 with Theorem 4.18 gives

Corollary 5.12. *Let Ω be a domain in $\mathbb{C}$ such that $\partial\Omega$ contains an angular point of order $\alpha \in [1, 2]$. Then*

$$\sup_{z \in \Omega} |\nabla(1/\lambda_\Omega(z))| \geq 2\alpha.$$

Our next aim is to examine the majorant in the Landau inequality. The following theorem shows that equality $M_n(z, \Delta, \Pi) = (n + |z|)(1 + |z|)^{n-2}$ occurs not only for the Koebe domain $H_2 - 1/4$ and its linear transformations.

Theorem 5.13. *Let $n \in \mathbb{N}$ and let Π be a simply connected domain in $\mathbb{C}$ such that there is an angular point w_0 on $\partial\Pi$ of order 2. Then, for any $z \in \Delta$,*

$$M_n(z, \Delta, \Pi) = (n + |z|)(1 + |z|)^{n-2}.$$

Proof. Clearly,

$$M_n(z, \Delta, \Pi) = M_n(|z|, \Delta, \Pi).$$

Hence, it is sufficient to consider the real values of z lying in $[0, 1)$ and to prove that

$$M_n(|z|, \Delta, \Pi) \geq M_n(|z|, \Delta, H_2) = (n + |z|)(1 + |z|)^{n-2}.$$

Also, we have to distinguish the cases of $w_0 \in \mathbb{C}$ and $w_0 = \infty$. Suppose first that $w_0 \neq \infty$. Without loss of generality, one can take $w_0 = 0$ and a domain Π satisfying

$$\Delta_2^+ \subset \Pi \subset \overline{\mathbb{C}} \setminus \Delta_0^-.$$

For $\gamma \in (1, +\infty)$ we consider the conformal mappings $\varphi_\gamma : \Delta \to \Delta_\alpha^+ \subset \Pi$ given by

$$\varphi_\gamma(z) = (\gamma - 1)^2 \left(\frac{1+z}{\gamma - z} \right)^2.$$

One has

$$\frac{1}{n!} \varphi_\gamma^{(n)}(z) = (\gamma - 1)^2 (\gamma + 1) \frac{(\gamma + 1)(n + 1 + 2(z - \gamma))}{(\gamma - z)^{n+2}}$$

and

$$\lambda_\Delta(z) = \frac{1}{1 - |z|^2}, \quad z \in \Delta, \quad \text{and} \quad \lambda_\Pi(x) = \frac{1}{4x + o(x)}, \quad \text{as } x \to 0^+.$$

Consequently, for any $x = |z| \in [0, 1)$,

$$M_n(z, \Delta, \Pi) \geq \lim_{\gamma \to 1^+} \frac{\lambda_\Pi(\varphi_\gamma(x))}{(\lambda_\Delta(x))^n} \frac{|\varphi_\gamma^{(n)}(x)|}{n!}$$

$$= \lim_{\gamma \to 1^+} \frac{(1 - x)^2 (1 - x^2)^n}{4(\gamma - 1)^2 (1 + x)^2} \frac{|\varphi_\gamma^{(n)}(x)|}{n!} = (1 + x)^{n-2}(n + x).$$

This completes the proof for $w_0 \neq \infty$.

Now, let $w_0 = \infty$. Without loss of generality we can suppose that

$$\frac{1}{\Delta_2^+} \subset \Pi \subset H_2 - 1.$$

As members of $A(\Delta, \Pi)$ we take functions g_γ defined by

$$g_\gamma(z) = \frac{1}{\varphi_\gamma(-z)} = (\gamma - 1)^{-2} \left(\frac{\gamma + z}{1 - z} \right)^2.$$

for any $\gamma \in (1, +\infty)$. One easily obtains the desired inequality using

$$\lim_{\gamma \to 1^+} (\gamma - 1)^2 \frac{|g_\gamma^{(n)}(x)|}{n!} = \frac{4(n + x)}{(1 - x)^{n+2}}.$$

The proof of Theorem 5.13 is complete. $\qquad\qquad\qquad\qquad\qquad\qquad\square$

One may generalize the proof of Theorem 5.13 in the following way. If $\partial\Pi$ contains an angular point $z_0 = 0$ (or $z_0 = \infty$) of order $\alpha \in [1, 2]$, then one can consider functions defined by

$$\varphi_{\gamma,\alpha}(z) = (\gamma - 1)^\alpha \left(\frac{1 - z}{\gamma + z} \right)^\alpha \quad \left(\text{or} \quad g_{\gamma,\alpha}(z) = (\gamma - 1)^{-\alpha} \left(\frac{\gamma + z}{1 - z} \right)^\alpha \right)$$

as members of $A(\Delta, \Pi)$ for any $\gamma \in (1, \infty)$. Since

$$\lim_{\gamma \to 1^+} \frac{|\varphi_{\gamma,\alpha}^{(n)}(x)|}{(\gamma - 1)^\alpha} = \lim_{\gamma \to 1^+} \frac{|g_{\gamma,\alpha}^{(n)}(x)|}{(\gamma - 1)^{-\alpha}} = \frac{d}{dx^n} \left(\frac{1 + x}{1 - x} \right)^\alpha, \quad x \in [0, 1),$$

one easily gets

$$M_n(x, \Delta, \Pi) \geq \frac{(1 + x)^{n-\alpha}(1 - x)^{n+\alpha}}{2\,\alpha\,n!} \frac{d}{dx^n} \left(\frac{1 + x}{1 - x} \right)^\alpha, \quad x \in [0, 1).$$

This consideration leads to the following theorem.

Theorem 5.14. *Let $n \in \mathbb{N}$ and let $z \in \Delta$. Suppose that Π is a domain in $\mathbb{C}$ such that $\partial\Pi$ contains an angular point of order $\alpha \in [1, 2]$. Then*

$$M_n(z, \Delta, \Pi) \geq M_n(z, \Delta, H_\alpha).$$

5.4   Domains in the extended complex plane

Let Ω be a hyperbolic simply connected domain in the extended complex plane $\overline{\mathbb{C}} = \mathbb{C} \cup \{\infty\}$. We consider the set $A(\Omega, \Pi)$ of $f : \Omega \to \Pi$, where Π is likewise a hyperbolic simply connected domain in $\overline{\mathbb{C}}$. Let λ_Ω and λ_Π be the densities of the Poincaré metric in Ω and Π with the Gaussian curvature $K = -4$. In the sequel we use the quantities $p \in (0, 1]$ and $q \in (0, 1]$ such that $D_\Delta(0, p)$ and $D_\Delta(0, q)$ equal $D_\Omega(z_0, \infty)$ and $D_\Pi(f(z_0), \infty)$, respectively. As above, by definition,

$$p := p(z_0) = \tanh D_\Omega(z_0, \infty), \quad q := q(f(z_0)) = \tanh D_\Pi(f(z_0), \infty). \quad (5.16)$$

Theorem 5.15 ([19]). *Suppose that Ω and Π are two hyperbolic simply connected proper subdomains of $\overline{\mathbb{C}}$, $K = -4$. For any $f \in A(\Omega, \Pi)$ the inequality*

$$\frac{1}{n!}|f^{(n)}(z_0)| \leq \left(q + \frac{1}{q} + p + \frac{1}{p} \right)^{n-1} \frac{(\lambda_\Omega(z_0))^n}{\lambda_\Pi(f(z_0))} \quad (5.17)$$

holds for any $n \geq 2$ and at any point $z_0 \in \Omega$. For each $n \geq 2$, each $p \in (0, 1]$, and each $q \in (0, 1]$ there exist $\Omega_0, \Pi_0, z_0 \in \Omega_0 \setminus \{\infty\}$ and $f_0 \in A(\Omega_0, \Pi_0)$ such that equality in (5.17) holds for them with $p = p(z_0)$ and $q = q(f(z_0))$.

Proof. For fixed $z_0 \in \Omega \setminus \{\infty\}$ we consider the conformal map Ψ_Ω of Ω onto Δ with $\Psi_\Omega(z_0) = 0$ and $\Psi'_\Omega(z_0) = \lambda_\Omega(z_0) > 0$. Obviously, this function has an expansion

$$\Psi_\Omega(z) = \lambda_\Omega(z_0) \left(z - z_0 + \sum_{k=2}^{\infty} A_k (z - z_0)^k \right), \tag{5.18}$$

valid in some neighbourhood of the point z_0. Let $b = \Psi_\Omega(\infty) \in \Delta$. Because of the conformal invariance of the hyperbolic distance, $D_\Delta(0, b)$ equals $D_\Omega(z_0, \infty)$. According to (5.16) we obtain $|b| = p = p(z_0)$. Let the function

$$\kappa(\zeta) = \frac{\zeta}{(1 + p\zeta)\left(1 + \frac{\zeta}{p}\right)}, \quad \zeta \in \Delta, \tag{5.19}$$

which is univalent in Δ, have an inverse function $K = \kappa^{-1}$ defined on $\kappa(\Delta)$. Obviously $\infty \in \kappa(\Delta)$ and $0 \in \kappa(\Delta)$. Using the expansion

$$K(z) = z + \sum_{k=2}^{\infty} B_k z^k,$$

valid in some neighbourhood of the origin, Baernstein and Schober (see Chapter 2, Theorem 2.15) showed that

$$|A_k| \leq B_k, \quad k \in \mathbb{N} \setminus \{1\}. \tag{5.20}$$

Consider now for each $f \in A(\Omega, \Pi)$ the function

$$g(\zeta) := f(\Phi_\Omega(\zeta)) = \sum_{k=0}^{\infty} a_k \zeta^k, \quad \zeta \in \Delta, \tag{5.21}$$

where Φ_Ω is inverse to Ψ_Ω. It is known by formula (4.5) that

$$\frac{1}{n!} \frac{f^{(n)}(z_0)}{(\lambda_\Omega(z_0))^n} = \sum_{k=1}^{n} A_{n,k}(z_0, \Omega) a_k, \tag{5.22}$$

where

$$A_{n,k}(z_0, \Omega) = \frac{(\lambda_\Omega(z_0, \Omega))^{-n}}{n!} \left((\Psi_\Omega(z))^k \right)^{(n)} \Big|_{z=z_0}. \tag{5.23}$$

The formulas (5.18) and (5.23) imply that $A_{n,n}(z_0, \Omega) \equiv 1$ and that the $A_{n,k}(z_0, \Omega)$, $k = 1, \ldots, n$, are polynomials with positive coefficients in $A_1, \ldots, A_{n-1}$. Hence, we get as a consequence of this fact and of the formulas (5.18)–(5.23) the inequalities

$$|A_{n,k}(z_0, \Omega)| \leq t_{n,k} := \frac{1}{n!} \left((K(z))^k \right)^{(n)} \Big|_{z=0}. \tag{5.24}$$

Thus,

$$\frac{1}{n!}\frac{|f^{(n)}(z_0)|}{(\lambda_\Omega(z_0))^n} \leq \sum_{k=1}^{n} t_{n,k}|a_k|, \tag{5.25}$$

where the positive quantities $t_{n,k}$ may be computed, according to the formulas (5.19), (5.24), and the Cauchy formula, as

$$t_{n,k} = \frac{1}{2\pi i}\int_{|\zeta|=r}\frac{(1+a\zeta+\zeta^2)^{n-1}(1-\zeta^2)}{\zeta^{n-k+1}}\,d\zeta, \quad a = \frac{1}{p}+p, \quad r \in (0,1). \tag{5.26}$$

Let $q = \tanh D_\Pi(f(z_0),\infty) \in (0,1]$. We need the expansion

$$\frac{\zeta}{(1-q\zeta)\left(1-\frac{\zeta}{q}\right)} = \zeta + \sum_{k=2}^{\infty}c_k(q)\zeta^k, \quad |\zeta| < q.$$

It is known that (see Chapter 2)

$$c_k(q) = \frac{\sum_{j=0}^{k-1}q^{2j}}{q^{k-1}} = \frac{q^k-\frac{1}{q^k}}{q-\frac{1}{q}}.$$

The properties

$$\frac{t_{n,k}}{c_k(q)} \geq \frac{t_{n,k+1}}{c_{k+1}(q)} > 0, \quad k = 1,\ldots,n-1, \tag{5.27}$$

and

$$\sum_{k=1}^{n}c_k(q)\,t_{n,k} = \left(a+q+\frac{1}{q}\right)^{n-1}, \quad a = \frac{1}{p}+p, \tag{5.28}$$

of the quantities $t_{n,k}$ will be proved below. Let us suppose for the moment that they are true. Applying the Cauchy inequality to (5.25) we get

$$\frac{1}{n!}\frac{|f^{(n)}(z_0)|}{(\lambda_\Omega(z_0))^n} \leq \left(\sum_{k=1}^{n}c_k(q)\,t_{n,k}\right)^{\frac{1}{2}}\left(\sum_{k=1}^{n}\frac{t_{n,k}}{c_k(q)}|a_k|^2\right)^{\frac{1}{2}}. \tag{5.29}$$

Let Φ_Π be the conformal map of Δ onto Π with $\Phi_\Pi(0) = a_0 = f(z_0)$ and $\Phi'_\Pi(0) = (\lambda_\Pi(f(z_0)))^{-1} > 0$. The function g defined in (5.21) is subordinate to the univalent function

$$\Phi_\Pi(\zeta) = a_0 + (\lambda_\Pi(f(z_0)))^{-1}\left(\zeta + \sum_{k=2}^{\infty}c_k\,\zeta^k\right).$$

The starting point for the rest of the proof is Theorem 2.5 which is a generalized version of the Rogosinski-Goluzin theorem.

The second ingredient at this point is Jenkins' theorem 2.7. We easily get

$$(\lambda_\Pi(f(z_0)))^2\sum_{k=1}^{n}\frac{t_{n,k}}{c_k(q)}|a_k|^2 \leq \sum_{k=1}^{n}\frac{t_{n,k}}{c_k(q)}|c_k|^2 \leq \sum_{k=1}^{n}c_k(q)\,t_{n,k}. \tag{5.30}$$

From (5.28), (5.29) and (5.30) we obtain the desired inequality (5.17).

To complete the proof of (5.17) we have to prove (5.27) and (5.28). An immediate consequence of (5.26) for $r \in (0, q)$ is

$$\sum_{k=1}^{n} c_k(q) \, t_{n,k} = \frac{1}{2\pi i} \int_{|\zeta|=r} \frac{(1 + a\zeta + \zeta^2)^{n-1}(1 - \zeta^2)}{\zeta^n} \sum_{k=1}^{n} c_k(q) \, \zeta^{k-1} \, d\zeta$$

$$= \frac{1}{2\pi i} \int_{|\zeta|=r} I(\zeta) \, d\zeta = \mathrm{Res}(I(\zeta), \zeta = 0)$$

$$= -\mathrm{Res}(I(\zeta), \zeta = q) - \mathrm{Res}(I(\zeta, \zeta = \frac{1}{q}) - \mathrm{Res}(I(\zeta), \zeta = \infty),$$

where

$$I(\zeta) = \frac{(1 - \zeta^2)}{(q - \zeta)\left(\frac{1}{q} - \zeta\right)} \frac{(1 + a\zeta + \zeta^2)^{n-1}}{\zeta^n}.$$

We observe that $\mathrm{Res}(I(\zeta), \zeta = 0) = \mathrm{Res}(I(\zeta), \zeta = \infty)$ and $\mathrm{Res}(I(\zeta), \zeta = q) = \mathrm{Res}(I(\zeta), \zeta = \frac{1}{q})$. Therefore

$$\sum_{k=1}^{n} c_k(q) \, t_{n,k} = -\mathrm{Res}(I(\zeta), \zeta = q) = \left(q + \frac{1}{q} + a\right)^{n-1}.$$

To prove (5.27) we consider the polynomial

$$P_n(\zeta) = (1 + a\zeta + \zeta^2)^n = \sum_{m=0}^{2n} b_m \zeta^m, \tag{5.31}$$

where, as before, $a = p + 1/p > 2$. The identity

$$P_n(\zeta) = (1 - \zeta^2) P_{n-1}(\zeta) + \zeta P_{n-1}(\zeta) P_1'(\zeta)$$

implies

$$P_n(\zeta) - \frac{\zeta}{n} P_n'(\zeta) = (1 - \zeta^2) P_{n-1}(\zeta).$$

Using (5.26) we identify $t_{n,k}$ as the coefficient of ζ^{n-k} in the polynomial on the right side of this equation. On the other hand, (5.31) shows that $b_{n-k} - \frac{n-k}{n} b_{n-k}$ is the coefficient of ζ^{n-k} in the polynomial on the left side. Therefore, we obtain

$$n \frac{t_{n,k}}{k} = b_{n-k}, \quad k = 1, \ldots, n.$$

This implies that the inequalities

$$\frac{t_{n,k}}{k} \geq \frac{t_{n,k+1}}{k+1} > 0, \quad k = 1, \ldots, n-1, \tag{5.32}$$

are equivalent to
$$1 \le b_1 \le b_2 \le \cdots \le b_{n-1},$$

which may be proved by mathematical induction on n considering the coefficients of the polynomials $(1 - \zeta) P_n(\zeta)$. Moreover, (5.32) implies the desired inequalities (5.27) because of the simple relations

$$\frac{c_{k+1}(q)}{c_k(q)} \ge \frac{k+1}{k}, \quad k = 1, \ldots, n - 1.$$

Now, it remains to prove that (5.17) is sharp. To this end we consider the following example. For any $n \in \mathbb{N}$ and any $a = p + 1/p > 2$, let

$$\Omega_0 = \overline{\mathbb{C}} \setminus [(a + 2)^{-1}, (a - 2)^{-1}],$$

$$f_0(z) = \frac{z}{1 - \left(a + q + \frac{1}{q}\right) z}$$

and

$$\Pi_0 = f_0(\Omega_0).$$

It is an easy task to derive that $\lambda_{\Omega_0}(0) = \lambda_{\Pi_0}(0) = 1$, that $\tanh D_{\Omega_0}(0, \infty)$ equals p, $\tanh D_{\Pi_0}(0, \infty)$ equals q and that

$$\frac{f_0^{(n)}(0)}{n!} = \left(p + \frac{1}{p} + q + \frac{1}{q}\right)^{n-1}.$$

This completes the proof of Theorem 5.15.   □

Consider now an improvement of (5.17) for convex domains.

If the domain Π is convex, then the coefficients $a_k, k \in \mathbb{N}$, of the functions g defined in (5.21) satisfy the inequalities of Rogosinski (see Theorem 4.12)

$$\lambda_\Pi(a_0)|a_k| \le 1, \quad k \in \mathbb{N}. \tag{5.33}$$

Therefore (5.25) and (5.33) imply

$$\frac{1}{n!} \left| f^{(n)}(z_0) \right| \le \frac{(\lambda_\Omega(z_0))^n}{\lambda_\Pi(f(z_0))} \sum_{k=1}^n t_{n,k}, \tag{5.34}$$

where we used the same definitions as in the proof of Theorem 5.15. A straightforward computation using (5.26) shows that

$$\sum_{k=1}^n t_{n,k} = \frac{1}{\pi} \int_0^\pi \left(p + \frac{1}{p} + 2\cos\theta\right)^{n-1} (1 + \cos\theta) \, d\theta =: C_n(p), \tag{5.35}$$

if $\infty \in \Omega$ and therefore $p \in (0, 1)$, respectively

$$\sum_{k=1}^{n} t_{n,k} = \binom{2n-1}{n}, \tag{5.36}$$

if $\infty \notin \Omega$. The formulas (5.34), (5.35) and (5.36) immediately yield

Theorem 5.16 ([19]). *Suppose that Ω is a hyperbolic simply connected proper subdomain of $\overline{\mathbb{C}}$ and that Π is a convex proper subdomain of $\mathbb{C}$, $K = -4$. Then the following estimates are valid:*

If $\infty \in \Omega$, then for any $f \in A(\Omega, \Pi)$ and any $z \in \Omega$ with hyperbolic distance $D_\Omega(z, \infty) = \text{Arctanh } p$, the inequality

$$\frac{1}{n!} |f^{(n)}(z)| \le C_n(p) \frac{(\lambda_\Omega(z))^n}{\lambda_\Pi(f(z))} \tag{5.37}$$

holds for any $n \in \mathbb{N}$.

Remark 5.17. The sharpness of (5.37) can be shown using Ω_0, H_1 and the function f_1 defined by

$$2f_1(z) := \frac{1 + K(z)}{1 - K(z)} = \sqrt{\frac{1 - (a-2)z}{1 - (a+2)z}}. \tag{5.38}$$

Indeed, the function f_1 maps Ω_0 conformally onto the half plane H_1, $K(0) = 0$ and $\lambda_{\Omega_0}(0) = \lambda_{\Pi_1}(1/2) = 1$. Now, we compare two expansions

$$f_1(z) = \frac{1}{2} + \sum_{k=1}^{\infty} (K(z))^k = \frac{1}{2} + \sum_{n=1}^{\infty} d_n z^n, \tag{5.39}$$

which are valid in some neighbourhood of the origin. The formulas (5.38), (5.39), and (5.24) immediately yield

$$d_n = \frac{1}{n!} \sum_{k=1}^{n} \left((K(z))^k \right)^{(n)} \bigg|_{z=0} = \sum_{k=1}^{n} t_{n,k} = C_n(p). \tag{5.40}$$

Thus,

$$\frac{1}{n!} \left| f_1^{(n)}(0) \right| = C_n(p) \frac{(\lambda_{\Omega_0}(0))^n}{\lambda_{\Pi_1}(f_1(0))}.$$

Moreover, using (5.39), (5.40) and the Taylor series of $(f_1)^2$ in some neighbourhood of the origin, we get

$$\left(\frac{1}{2} + \sum_{k=1}^{\infty} C_k(p) z^k \right)^2 = \frac{1}{4} + \sum_{n=1}^{\infty} (a+2)^{n-1} z^n,$$

which implies

$$C_1(p) = 1, \quad C_2(p) = a + 1 = p + \frac{1}{p} + 1, \text{ and}$$

$$C_n(p) = \left(p + \frac{1}{p} + 2\right)^{n-1} - \sum_{k=1}^{n-2} C_k(p)C_{n-k}(p), \quad n \geq 3.$$

These formulas show the difference between the bounds in (5.17) and (5.37).

Remark 5.18. It may be worthwhile to notice that the direct determination of the integral (5.35) and the computation of the Taylor coefficients of the function (5.38) with the help of binomial series can be exploited to prove relations between binomial coefficients that seem to be not known. We were not able to find another method to prove these identities than the geometrical method above. For the convenience of the interested reader we mention such a formula. For $m = 0, \ldots, n-1$, $n \in \mathbb{N}$, one may get in this way the identities

$$(-1)^n \binom{n-1}{m} \binom{m}{\left[\frac{m}{2}\right]}$$

$$= 2^m \sum_{k=0}^{n} \binom{1/2}{k} \binom{-1/2}{n-k} \left(\sum_j (-1)^j \binom{k}{j} \binom{n-k}{m+1-j}\right),$$

where j varies in the second sum in the range determined by the binomial coefficients.

5.5   Maps from convex into concave domains

We shall consider the case when $\infty \in \Pi$ and the punishing factor is a function of the hyperbolic distance between $f(z)$ and ∞. Moreover, we suppose that $\Pi = \overline{\mathbb{C}} \setminus \Pi_1$, where Π_1 is a compact convex subset of $\mathbb{C}$ containing more than one point. Domains Π of this type will be called concave domains.

In the sequel we use the quantity $p \in (0, 1)$ such that

$$D_\Delta(0, p) = \frac{1}{2} \log \frac{1+p}{1-p}$$

equals $D_\Pi(f(z_0), \infty)$. This means that

$$p := p(f(z_0)) = \tanh D_\Pi(f(z_0), \infty), \tag{5.41}$$

where, as usual, $\tanh x = (e^x - e^{-x})/(e^x + e^{-x})$.

A central part in the proofs is played by functions h satisfying the following conditions:

(i) The function h is meromorphic and injective on Δ and h has its pole at a point $p \in (0, 1)$.

(ii) The set $\overline{\mathbb{C}} \setminus h(\Delta)$ is convex.

(iii) $h(0) = h'(0) - 1 = 0$.

We called functions with the properties (i)–(iii) concave univalent functions with pole p and for the family of those functions we used the abbreviation $Co(p)$. For older results on such functions compare [123], [117] and [108] and for newer ones [30], [31], [15], [20], [23], [166] and [167].

Theorem 5.19 ([28]). *Let Π be a concave domain. Then for all $n \in \mathbb{N}$, $f \in A(\Delta, \Pi)$, $z_0 \in \Delta$, $f(z_0)$ finite, and p as in (5.41) the inequalities*

$$\frac{|f^{(n)}(z_0)|}{n!} \frac{(1 - |z_0|^2)^n}{R_\Pi(f(z_0))} \leq \frac{1}{1 - p^2} \left(\left(|z_0| + \frac{1}{p} \right)^{n-1} - p^2(|z_0| + p)^{n-1} \right) \quad (5.42)$$

are valid. Equality in (5.42) is attained if $z_0 = r \in [0, 1)$, and $f = E \circ T$,

$$E(z) = \frac{z}{(1 - zp)(1 - z/p)} \quad \text{and} \quad T(z) = \frac{z - r}{1 - zr}, \quad z \in \Delta.$$

These functions f are conformal maps of Δ onto the domain

$$\overline{\mathbb{C}} \setminus \left[\frac{-p}{(1 - p)^2}, \frac{-p}{(1 + p)^2} \right].$$

Corollary 5.20 (compare [31]). *Let Π be a concave domain. Then for all $n \geq 2$, $f \in A(\Delta, \Pi)$, $z_0 \in \Delta$, $f(z_0)$ finite, and p as in (5.41) the inequalities*

$$\frac{|f^{(n)}(z_0)|}{n!} \frac{(1 - |z_0|^2)^n}{R_\Pi(f(z_0))} \leq \frac{(1 + p)^{n-2}}{p^{n-1}} \sum_{k=0}^{n} p^k \quad (5.43)$$

are valid. The constant on the right side cannot be replaced by a smaller one independent of Π, f, z_0, and $f(z_0)$.

Proof of Theorem 5.19. We consider the function $g \in A(\Delta, \Pi)$ defined by

$$g(\zeta) = f\left(\frac{\zeta + z_0}{1 + \overline{z_0}\zeta} \right), \quad \zeta \in \Delta.$$

Since $f(z_0)$ is finite, the function g has a Taylor expansion

$$g(\zeta) = f(z_0) + \sum_{k=1}^{\infty} a_k \zeta^k$$

valid on some neighbourhood of the origin. It is clear that, under our circumstances, we may also use the identity (4.3), which has been proved for functions holomorphic in the unit disc.

On the other hand, we use that $g(0) = f(z_0)$ and $g(\Delta) \subset \Pi$. If we let, for this proof, $\Phi := \Phi_{\Pi, f(z_0)}$ be defined as above, we see that there exists a function $\omega_1 : \Delta \to \overline{\Delta}$ holomorphic on Δ such that

$$g(\zeta) = \Phi(\zeta \omega_1(\zeta)), \quad \zeta \in \Delta. \tag{5.44}$$

Further, we conclude from the definition $p = \tanh D_\Pi(f(z_0), \infty)$ that there exists $\varphi \in [0, 2\pi]$ such that $\Phi(pe^{i\varphi}) = \infty$. Now, we consider the function h defined on Δ by

$$h(\zeta) = \frac{e^{-i\varphi}}{R_\Pi(f(z_0))}(\Phi(e^{i\varphi}\zeta) - f(z_0)), \quad \zeta \in \Delta. \tag{5.45}$$

This function h belongs to the class $Co(p)$ and from (5.44) and (5.45) we see that there exists a function $h \in Co(p)$ and a function $\omega = e^{-i\varphi}\omega_1 : \Delta \to \overline{\Delta}$ such that

$$g(\zeta) = e^{i\varphi}h(\zeta \omega(\zeta))R_\Pi(f(z_0)) + f(z_0), \quad \zeta \in \Delta. \tag{5.46}$$

In Theorem 8.4 below (compare [26]) we will consider $h \in Co(p), \omega$ as above such that the Taylor expansion

$$h(\zeta \omega(\zeta)) = \sum_{k=1}^{\infty} \alpha_k \zeta^k \tag{5.47}$$

is valid on some neighbourhood of the origin. We will prove there that for $k \in \mathbb{N}$ the inequalities

$$|\alpha_k| \le \frac{1 - p^{2k}}{(1 - p^2)p^{k-1}} = \frac{1}{1 - p^2}\left(\frac{1}{p^{k-1}} - p^2 p^{k-1}\right) \tag{5.48}$$

are valid. After these preparations (5.42) is an immediate consequence of (4.3), (5.46), (5.47), and (5.48).

The second assertion of Theorem 5.19 becomes clear from the above proof and the expansion

$$E(z) = \sum_{k=1}^{\infty} \frac{1 - p^{2k}}{(1 - p^2)p^{k-1}}z^k, \quad |z| < p.$$

This concludes the proof of Theorem 5.19. □

Theorem 5.21 ([28]). *Let $\Omega \subset \mathbb{C}$ be a convex proper subdomain of $\mathbb{C}$. Let further Π be a concave subdomain of $\overline{\mathbb{C}}$ and $f : \Omega \to \Pi$ be meromorphic on Ω. Then for all $n \ge 2$, $z_0 \in \Omega$, and $f(z_0)$ finite, and p as in (5.41) the inequalities*

$$\frac{|f^{(n)}(z_0)|}{n!}\frac{(R_\Omega(z_0))^n}{R_\Pi(f(z_0))} \le \frac{(1 + p)^{n-2}}{p^{n-1}}\sum_{k=0}^{n}p^k \tag{5.49}$$

are valid. Equality in (5.49) is attained for

$$\Omega = \{z \mid \mathrm{Re}\, z < 1/2\} \quad and \quad \Pi = \overline{\mathbb{C}} \setminus \left[\frac{-p}{(1-p)^2}, \frac{-p}{(1+p)^2} \right],$$

$z_0 = f(z_0) = 0$, *where the function*

$$f(z) = E\left(\frac{z}{1-z} \right) = \frac{z(1-z)}{(1 - z(1+p))(1 - z(1+p)/p)}$$

maps Ω *onto* Π *conformally.*

Proof. Let $n \geq 2$. According to the central formula (4.5) we have to prove that

$$\left| \sum_{k=1}^{n} c_k A_{n,k} \right| \leq \frac{(1+p)^{n-2}}{p^{n-1}} \sum_{k=0}^{n} p^k,$$

where c_k are the Taylor coefficients of a function $h \in Co(p)$, and, as usual, $A_{n,k}$ is the n-th coefficient in the Taylor expansion of F^k for the function F inverse to an arbitrary function $s \in K$. Because of the Schur-Jabotinsky theorem (compare the proof of Theorem 4.15), we have to prove that

$$\left| \sum_{l=0}^{n-1} \frac{n-l}{n} c_{n-l}\, a_{l,n} \right| \leq \frac{(1+p)^{n-2}}{p^{n-1}} \sum_{k=0}^{n} p^k. \tag{5.50}$$

Using the Marx-Strohhäcker inequality as above, we get the following formula for the sum appearing in (5.50):

$$\sum_{l=0}^{n-1} \frac{n-l}{n} c_{n-l}\, a_{l,n} = c_n + \sum_{\sigma=1}^{n-1} \frac{1}{n} \binom{n}{\sigma} \sum_{j=\sigma}^{n-1} (n-j) c_{n-j} d_{j-\sigma,\sigma}. \tag{5.51}$$

Now it is easily verified that it is sufficient to obtain the inequality

$$\left| \sum_{j=\sigma}^{n-1} (n-j) c_{n-j} d_{j-\sigma} \right| \leq \frac{(n-\sigma)(1 - p^{2(n-\sigma)})}{p^{n-\sigma-1}(1 - p^2)},$$

where $d_{j-\sigma}$ are the coefficients of a unimodular bounded holomorphic function in the unit disc as in the proof of Theorem 4.15. The latter inequality is equivalent to

$$\left| \sum_{\tau=0}^{q-1} (q-\tau)\, c_{q-\tau} d_\tau \right| \leq \frac{q(1 - p^{2q})}{p^{q-1}(1 - p^2)}. \tag{5.52}$$

As is proved in [20] and [167], for any $h \in Co(p)$ there exists a function $v_1 : \Delta \to \overline{\Delta}$ such that

$$h(z) = \frac{z}{1 - z/p} \left(1 + \frac{v_1(z) pz}{1 - zp} \right), \quad z \in \Delta.$$

Using this and the generalized version of Sheill-Small's Lemma 2.10 (see Remark 2.12) in the limiting case $r \to 1$, i.e., the inequality

$$|P * \tilde{h}(e^{i\theta})| \le \max_{|z|=1} |P(z)| \left(\sum_{k=0}^{q-1} |u_k|^2 \right)^{1/2} \left(\sum_{k=0}^{q-1} |v_k|^2 \right)^{1/2}$$

for the polynomial

$$P(z) = \sum_{\tau=0}^{q-1} d_\tau (q-\tau) z^{q-\tau}$$

and the functions defined by

$$w_i(z) = \omega(z), \ i = 1, 2, 3, \quad z \in \Delta,$$

and

$$U(z) = \sum_{k=0}^{q-1} \left(\frac{z}{p} \right)^k \quad \text{and} \quad V(z) = 1 + \frac{v_1(z)pz}{1 - zp} = \sum_{k=0}^{\infty} v_k z^k, \quad z \in \Delta,$$

we obtain the inequality

$$\left| \sum_{\tau=0}^{q-1} (q-\tau) c_{q-\tau} d_\tau e^{i(q-\tau)\theta} \right| \le \max_{|z|=1} |P(z)| \left(\sum_{k=0}^{q-1} p^{-2k} \right)^{1/2} \left(\sum_{k=0}^{q-1} |v_k|^2 \right)^{1/2}.$$

To estimate the last factor of the right side in this inequality we use that the function $V(z) - 1$ is quasi-subordinated to the function $pz/(1 - zp)$. According to Theorem 2.4 we get

$$\sum_{k=0}^{q-1} |v_k|^2 \le \sum_{k=0}^{q-1} p^{2k}.$$

Hence,

$$\left| \sum_{\tau=0}^{q-1} (q-\tau) c_{q-\tau} d_\tau \right| \le \max_{|z|=1} |P(z)| \left(\sum_{k=0}^{q-1} p^{-2k} \right)^{1/2} \left(\sum_{k=0}^{q-1} p^{2k} \right)^{1/2}$$

$$= \max_{|z|=1} |P(z)| \frac{1 - p^{2q}}{p^{q-1}(1 - p^2)}.$$

Fejér's inequality (see [61] and [159] or apply the theory of linear functionals on H^∞ in chapter 8 of [59])

$$\left| \sum_{\tau=0}^{q-1} (q-\tau) d_\tau \right| \le q$$

shows that

$$\left| P\left(e^{-i\theta} \right) \right| \le q, \quad \theta \in [0, 2\pi].$$

This concludes the proof of the desired inequality.

The extremal property of the function f mentioned in the second assertion of Theorem 5.21 follows directly from the computation of the Taylor coefficients of its expansion with expansion point at the origin.

The proof of Theorem 5.21 is complete. $\qquad\qquad\square$

Chapter 6

Multiply connected domains

In the preceding chapters we considered punishing factors for simply connected domains, except the case $C_2(\Omega, \Pi)$. Namely, in Section 4.6 it was proved that for all hyperbolic domains $\Omega \subset \overline{\mathbb{C}}$ and $\Pi \subset \overline{\mathbb{C}}$,

$$
C_2(\Omega, \Pi) = C_2(\Pi, \Omega) = \frac{1}{2} \left(\sup_{z \in \Omega} |\nabla (1/\lambda_\Omega(z))| + \sup_{w \in \Pi} |\nabla (1/\lambda_\Pi(w))| \right).
$$

In particular, it was proved that

$$
C_2(A_1, A_2) = \sqrt{1 + 4M_1^2} + \sqrt{1 + 4M_2^2},
$$

where A_1 and A_2 are annuli with moduli M_1 and M_2, respectively.

The main aim of this chapter is to describe multiply connected domains by properties that guarantee existence of finite punishing factors for all n. Also we will define finite modified punishing factors and consider some examples.

6.1 Finitely connected domains

The theorems of Chapters 4 and 5 show that one cannot expect that

$$
C_n(\Omega, \Pi) := \sup\{M_n(z, \Omega, \Pi) \mid z \in \Omega\} \tag{6.1}
$$

is always finite. The central existence theorem for finitely connected domains is the following assertion.

Theorem 6.1 ([21]). *Let Ω and Π be finitely connected hyperbolic domains in $\mathbb{C}$. Then the punishing factors $C_n(\Omega, \Pi)$ are finite for all $n \in \mathbb{N}$ if and only if both $\partial\Omega$ and $\partial\Pi$ do not contain isolated points.*

Firstly, we prove some necessary lemmas and propositions. We shall use the hyperbolic radius $R(z, \Pi)$ which is reciprocal to the density of hyperbolic metric with Gaussian curvature $K = -4$.

Let Π be an m-connected hyperbolic domain in $\overline{\mathbb{C}}$ where $m \geq 2$, i.e., $\partial \Pi$ consists of m disjoint connected sets Γ_k, $k = 1, \ldots, m$. We shall consider m associated simply connected domains $\Pi_k \subset \overline{\mathbb{C}}$, $k = 1, \ldots, m$, defined by the relations

$$\partial \Pi_k = \Gamma_k \quad \text{and} \quad \Pi \subset \Pi_k, \quad k = 1, \ldots, m.$$

It is clear that $\Gamma_j \subset \Pi_k$ for $j \neq k$ and that

$$\Pi = \bigcap_{k=1}^{m} \Pi_k.$$

Naturally, the connected sets Γ_k are either points or continua. It is decisive for our proofs that the behaviour of the hyperbolic radius of Π and Π_k is the same at the common boundary. If Γ_k is a point, this is known (see [3] and [55]). We will use the formula for this case later on (compare (6.7) and 6.8)). In the case that Γ_k is a continuum, this is the content of the following lemma.

Lemma 6.2. *Let Π and Π_k be as above. If $\Gamma_k = \partial \Pi_k$ is not a point, then for $z_0 \in \Gamma_k$,*

$$\lim_{z \to z_0, \, z \in \Pi} \frac{R(z, \Pi)}{R(z, \Pi_k)} = 1. \tag{6.2}$$

Proof. Since Γ_k is a continuum, Π_k is a simply connected domain in $\overline{\mathbb{C}}$. Riemann's mapping theorem implies that there exists a conformal map Φ of Δ onto Π_k. The inverse function Φ^{-1} maps Π onto a set $\Phi^{-1}(\Pi) \subset \Delta$. For all sufficiently small positive numbers ϵ the relations

$$D(\epsilon) := \{w \mid e^{-\epsilon \pi} < |w| < 1\} \subset \Phi^{-1}(\Pi) \subset \Delta$$

are valid. According to Theorem 3.5 they imply

$$R(w, D(\epsilon)) \leq R(w, \Phi^{-1}(\Pi)) \leq R(w, \Delta), \quad w \in D(\epsilon). \tag{6.3}$$

Combining (6.3) and the conformal invariance of the hyperbolic metric,

$$\frac{|dz|}{R(z, \Pi)} = \frac{|dw|}{R(w, \Phi^{-1}(\Pi))}, \quad \text{and} \quad \frac{|dz|}{R(z, \Pi_k)} = \frac{|dw|}{R(w, \Delta)},$$

we get, for any $w = \Phi^{-1}(z) \in D(\epsilon)$,

$$1 \geq \frac{R(z, \Pi)}{R(z, \Pi_k)} = \frac{R(w, \Phi^{-1}(\Pi))}{R(w, \Delta)} \geq \frac{R(w, D(\epsilon))}{R(w, \Delta)}. \tag{6.4}$$

The use of

$$R(w, \Delta) = 1 - |w|^2 \quad \text{and} \quad R(w, D(\epsilon)) = 2\epsilon |w| \sin\left(\frac{1}{\epsilon} \log \frac{1}{|w|}\right)$$

and (6.4) yields

$$\lim_{|w|\to 1-} \frac{R(w,\Phi^{-1}(\Pi))}{R(w,\Delta)} = \lim_{|w|\to 1-} \frac{R(w,D(\epsilon))}{1-|w|^2} = 1. \tag{6.5}$$

Since $|\Phi^{-1}(z)| \to 1^-$ as $z \to z_0$, the assertion (6.2) of Lemma 6.2 follows from (6.4) and (6.5). $\qquad\square$

The second preparation for the proof of Theorem 6.1 is concerned with the Taylor coefficients $a_n(f)$ of the local expansion

$$f(z) = \sum_{n=0}^{\infty} a_n(f)z^n$$

for a function $f \in A(\Delta,\Pi)$ that is holomorphic at the origin. We wish to describe now all finitely connected hyperbolic domains for which the quantities

$$A_n(\Pi) := \sup_{f\in A(\Delta,\Pi)} \frac{|a_n(f)|}{R(a_0(f),\Pi)}$$

are finite for all $n \in \mathbb{N}$. One might expect that boundedness of Π is the decisive condition, but the proof of the following proposition reveals that this is not the case.

Lemma 6.3. *Let Π be a finitely connected hyperbolic domain in $\mathbb{C}$. Then the following statements are equivalent.*

(a) *$\partial\Pi$ does not contain isolated points.*

(b) *$A_n(\Pi)$ is finite for any $n \in \mathbb{N}$.*

(c) *$C_n(\Delta,\Pi)$ is finite for any $n \in \mathbb{N}$.*

(d) *There exists a constant K_Π such that, for all $w \in \Pi$,*

$$\operatorname{dist}(w,\partial\Pi) \le R(w,\Pi) \le K_\Pi \operatorname{dist}(w,\partial\Pi).$$

Proof. If Π is a simply connected domain, then Lemma 6.3 is a consequence of known facts (see [55]).

It is also known that (a)$\Leftrightarrow$(d) is valid (see [130] and [77]).

Suppose now that Π is m-connected with $m \ge 2$. We will prove that (a) $\Rightarrow$ (b) $\Rightarrow$ (c) $\Rightarrow$ (d) $\Rightarrow$ (a).

(a)$\Rightarrow$(b). The condition (a) means that the associated domains Π_k, $k = 1,\dots,m$, are hyperbolic. Since $\Pi \subset \mathbb{C}$, we may suppose, without loss of generality, that

$$\Pi_1 \subset \mathbb{C}, \quad \text{and} \quad \infty \in \Pi_k \subset \overline{\mathbb{C}}, \quad k = 2,\dots,m.$$

Since $f \in A(\Delta, \Pi) \subset A(\Delta, \Pi_k)$, we can apply theorems of Chapter 5 to get

$$A_n(\Pi) \le \sup\{A(w, n) \mid w \in \Pi\},$$

where $n \in \mathbb{N}$ and

$$A(w, n) := \frac{\min\{\sigma_k(w) \, R(w, \Pi_k) \mid k = 1, \ldots, m\}}{R(w, \Pi)} \tag{6.6}$$

with $\sigma_1(w) \equiv n$ and

$$\sigma_k(w) = 4^{n-1}(1 - \exp(-4\delta_k(w)))^{1-n}, \quad k = 2, \ldots, m.$$

In the latter formula the quantity $\delta_k(w) = \delta(w, \infty, \Pi_k)$ means the hyperbolic distance from w to ∞ in the domain Π_k.

The function $A(\cdot, n)$ is positive and continuous on Π, since hyperbolic radii and the functions σ_k are positive and real-analytic functions on Π.

Suppose that $A_n(\Pi) = \infty$ for one number $n \in \mathbb{N}$. Then there exists a sequence of points $w_j \in \Pi, j \in \mathbb{N}$, such that

$$w_j \to w_0 \in \partial\Pi, \quad \text{and} \quad A(w_j, n) \to \infty \quad \text{as} \quad j \to \infty.$$

But this is impossible. Actually, if $w_0 \in \partial\Pi_1$, then

$$\lim_{j \to \infty} A(w_j, n) \le n \lim_{j \to \infty} \frac{R(w_j, \Pi_1)}{R(w_j, \Pi)} = n$$

according to (6.6) and Lemma 6.2. If $w_0 \in \partial\Pi_k$ for some $k \in \{2, \ldots, m\}$, then

$$\lim_{j \to \infty} A(w_j, n) \le 4^{n-1} \lim_{j \to \infty} (1 - \exp(-4\delta_k(w_j)))^{1-n} \frac{R(w_j, \Pi_k)}{R(w_j, \Pi)} = 4^{n-1}$$

according to (6.6), Lemma 6.2 and the equation

$$\lim_{w \to w_0} \delta_k(w) = \delta(w, \infty, \Pi_k) = \infty.$$

Thus, our assumption leads to a contradiction and therefore (b) holds.

(b)$\Rightarrow$(c). Let $f \in A(\Delta, \Pi)$. Then, for any fixed $z \in \Delta$, the function

$$g : \Delta \to \Pi, \zeta \mapsto g(\zeta) = f\left(\frac{\zeta + z}{1 - \overline{z}\zeta}\right),$$

also belongs to $A(\Delta, \Pi)$. Using the formula (4.3)

$$\frac{f^{(n)}(z)(1 - |z|^2)^n}{n!} = \sum_{k=1}^{n} \binom{n-1}{n-k} \overline{z}^{n-k} a_k(g)$$

and the equation $f(z) = a_0(g)$, we obtain

$$C_n(\Delta, \Pi) \leq 2^{n-1} \max_{1 \leq k \leq n} A_k(\Pi).$$

This implies the desired implication.

(c)$\Rightarrow$(d). Let $\Phi := \Phi_\Pi$ be a covering map of Δ onto Π. On the one hand, $\Phi \in A(\Delta, \Pi)$. Hence, the validity of (c) implies that for all $z \in \Delta$,

$$\frac{(1 - |z|^2)^2 |\Phi''(z)|}{R(\Phi(z), \Pi)} \leq 2 C_2(\Delta, \Pi) < \infty.$$

On the other hand, $R(w, \Pi) = |\Phi'(z)|(1 - |z|^2)$ for $w = \Phi(z)$, $z \in \Delta$. Straightforward computations yield that

$$|\nabla R(w, \Pi)| = 2 \left| \frac{\partial R(w, \Pi)}{\partial w} \right| = \left| (1 - |z|^2) \frac{\Phi''(z)}{\Phi'(z)} - 2\bar{z} \right|$$

$$= \left| \frac{(1 - |z|^2)^2 |\Phi''(z)|}{R(\Phi(z), \Pi)} \frac{|\Phi'(z)|}{\Phi'(z)} - 2\bar{z} \right|.$$

Therefore

$$|\nabla R(w, \Pi)| \leq 2 + 2 C_2(\Delta, \Pi), \quad w \in \Pi.$$

Since $R(w, \Pi) = 0$ for $w \in \partial\Pi$, we immediately get

$$R(w, \Pi) \leq (2 + 2 C_2(\Delta, \Pi)) \, \mathrm{dist}(w, \partial\Pi), \quad w \in \Pi.$$

This proves the right inequality. The left inequality in (d) is true for any hyperbolic domain (see [3] and [55]). This completes the proof of (c)$\Rightarrow$(d).

The implication (d)$\Rightarrow$(a) is a simple consequence of known facts. Actually, if $\partial\Pi$ contains an isolated point w_0, then (see [3] and [55])

$$R(w, \Pi) = (2 + o(1))|w - w_0| \log \frac{1}{|w - w_0|}, \quad w \to w_0 \in \mathbb{C}, \tag{6.7}$$

and

$$R(w, \Pi) = (2 + o(1))|w| \log |w|, \quad w \to w_0 = \infty. \tag{6.8}$$

Hence, (d) cannot hold if $\partial\Pi$ contains an isolated point.

This completes the proof of Lemma 6.3. $\qquad\square$

Proof of Theorem 6.1. Suppose that $\partial\Omega$ and $\partial\Pi$ do not contain isolated points and that $f \in A(\Omega, \Pi)$.

For any fixed $z \in \Omega$ we consider the function

$$g : \Delta \to \Pi, \ \zeta \mapsto g(\zeta) = f(z + \rho\zeta),$$

where $\rho = \mathrm{dist}(z, \partial\Omega)$. It it evident that $g \in A(\Delta, \Omega)$ and that

$$a_0(g) = g(0) = f(z) \quad \text{and} \quad a_n(g) = \rho^n f^{(n)}(z), \; n \in \mathbb{N}.$$

Since $\partial\Pi$ does not contain isolated points, Lemma 6.3 implies that the quantities $A_n(\Pi)$ are finite for all $n \in \mathbb{N}$. Hence,

$$\frac{(\mathrm{dist}(z, \partial\Omega))^n |f^{(n)}(z)|}{R(f(z), \Pi)} = \frac{|a_n(g)|}{R(a_0(g), \Pi)} \le A_n(\Pi), \qquad (6.9)$$

for all $n \in \mathbb{N}$. Using the equivalence (a)$\Leftrightarrow$(d) of Lemma 6.3 with respect to Ω, we get

$$R(z, \Omega) \le K_\Omega \, \mathrm{dist}(z, \partial\Omega), \quad z \in \Omega,$$

for a positive constant K_Ω. This inequality and (6.9) imply that

$$|f^{(n)}(z)| \le A_n(\Pi)(K_\Omega)^n \frac{R(f(z), \Pi)}{(R(z, \Omega))^n}, \quad z \in \Omega.$$

Thus,

$$C_n(\Omega, \Pi) \le \frac{A_n(\Pi)(K_\Omega)^n}{n!} < \infty,$$

for all $n \in \mathbb{N}$. This completes the proof of one direction of the assertion.

Suppose now that Ω and Π are finitely connected hyperbolic domains in $\mathbb{C}$ and that $C_2(\Omega, \Pi)$ is finite. Firstly, we use some counterexamples to show that $\partial\Pi$ cannot have isolated points. Suppose on the contrary that $\partial\Pi$ has at least one isolated point w_0. Since $C_n(\Omega, \Pi)$ is invariant under linear transformations of Ω and Π, we may restrict ourselves to consideration of the following two cases without loss of generality. Either

$$w_0 = 0 \in \partial\Pi \quad \text{and} \quad \Delta' = \Delta \setminus \{0\} \subset \Pi \qquad (6.10)$$

or

$$w_0 = \infty \in \partial\Pi \quad \text{and} \quad D_\infty = \{w \mid 1 < |w| < \infty\} \subset \Pi. \qquad (6.11)$$

In the case (6.10), we consider the functions

$$\Psi_t := f_t \circ \Phi_\Omega^{-1} \in A(\Omega, \Delta') \subset A(\Omega, \Pi), \quad t \in (0, \infty),$$

where Φ_Ω^{-1} is the inverse of a covering map Φ_Ω of Δ onto Ω and f_t is defined by

$$f_t : \Delta \to \Delta', \; \zeta \mapsto f_t(\zeta) = \exp\left(-t\,\frac{1+\zeta}{1-\zeta}\right).$$

On the one hand, at $z_0 = \Phi_\Omega(0)$ the condition $C_2(\Omega, \Pi) < \infty$ implies

$$\frac{|\Psi_t''(z_0)|\,(R(z_0, \Omega))^2}{2\,C_2(\Omega, \Pi)} \le R(\Psi_t(z_0), \Pi) = R(e^{-t}, \Pi) = (2 + o(1))e^{-t}t$$

as $t \to \infty$.

On the other hand, straightforward computations show that this is impossible, since

$$|\Psi_t''(z_0)|\,(R(z_0, \Omega))^2 = 4\,e^{-t}\left|t^2 - t + t\,\frac{\Phi_\Omega''(0)}{\Phi_\Omega'(0)}\right|. \tag{6.12}$$

In the case (6.11) we examine the functions

$$\tilde{\Psi}_t := \tilde{f}_t \circ \Phi_\Omega^{-1} \in A(\Omega, D_\infty) \subset A(\Omega, \Pi), \quad t \in (0, \infty),$$

where Φ_Ω^{-1} is defined as above and

$$\tilde{f}_t : \Delta \to D_\infty, \; \zeta \mapsto \tilde{f}_t(\zeta) = \exp\left(t\,\frac{1+\zeta}{1-\zeta}\right).$$

In this case, we get for $z_0 = \Phi_\Omega(0)$,

$$\left|\tilde{\Psi}_t''(z_0)\right|(R(z_0, \Omega))^2 = e^t\,t^2(4 + o(1)),$$

as $t \to \infty$. This contradicts the condition $C_2(\Omega, \Pi) < \infty$ which implies

$$\frac{\left|\tilde{\Psi}_t''(z_0)\right|(R(z_0, \Omega))^2}{2\,C_2(\Omega, \Pi)} \le R(e^t, \Pi) = (2 + o(1))e^t\,t,$$

as $t \to \infty$. Thus, $\partial\Pi$ does not contain isolated points.

Now, we prove that the analogous assertion holds true for Ω. Firstly, we have just proved that there exists a positive constant K_Π such that

$$|\nabla R(w, \Pi)| \le K_\Pi, \quad w \in \Pi.$$

Now, let h be a covering map of Ω onto Π defined by

$$h = \Phi_\Pi \circ \Phi_\Omega^{-1} \in A(\Omega, \Pi).$$

Computations using the equality $R(h(z), \Pi) = |h'(z)|R(z, \Omega)$ show that

$$\frac{h''(z)(R(z, \Omega))^2}{2\,R(w, \Pi)} = \frac{(h'(z))^2}{|h'(z)|^2}\,\frac{\partial R(w, \Pi)}{\partial w} - \frac{h'(z)}{|h'(z)|}\,\frac{\partial R(z, \Omega)}{\partial z},$$

where $w = h(z), z \in \Omega$. It follows that

$$|\nabla R(z, \Omega)| \le |\nabla R(w, \Pi)| + \frac{|h''(z)|(R(z, \Omega))^2}{2\,R(w, \Pi)} \le K_\Pi + C_2(\Omega, \Pi)$$

for any $z \in \Omega$. The boundedness of $|\nabla R(\cdot, \Omega)|$ on Ω implies that $\partial\Omega$ has no isolated point.

This completes the proof of Theorem 6.1. $\qquad\qquad\square$

6.2 Pairs of arbitrary domains

We shall show that it is possible to define finite modified punishing factors even if Ω contains the point at infinity and $\partial\Omega$ contains isolated points. To this end, we use the Euclidean distance, $\mathrm{dist}(w, \partial\Omega)$, from a point to the boundary of a domain.

As a modification of Theorem 6.1 we get

Theorem 6.4 ([21]). *Let Ω be an arbitrary hyperbolic domain in $\overline{\mathbb{C}}$ and Π a finitely connected hyperbolic domain in $\mathbb{C}$. Then the modified punishing factors*

$$\tilde{C}_n(\Omega, \Pi) := \sup_{f \in A(\Omega, \Pi)} \sup_{z \in \Omega} \frac{|f^{(n)}(z)|(\mathrm{dist}(z, \partial\Omega))^n}{n!\,\mathrm{dist}(f(z), \partial\Pi)}$$

are finite for all $n \in \mathbb{N}$ if and only if $\partial\Pi$ does not contain isolated points.

Proof. Let $f \in A(\Omega, \Pi)$. If Π is a finitely connected hyperbolic domain in $\mathbb{C}$ and $\partial\Pi$ does not contain isolated points, then the inequality (6.9) holds for any $n \in \mathbb{N}$. Moreover, by the statement (d) of Lemma 6.3, there exists a positive constant K_Π such that

$$R(f(z), \Pi) \leq K_\Pi \mathrm{dist}(f(z), \partial\Pi), \quad z \in \Omega.$$

This inequality and (6.9) imply that

$$n!\,\tilde{C}_n(\Omega, \Pi) \leq A_n(\Pi)\,K_\Pi, \quad n \in \mathbb{N}.$$

This proves one direction of the assertion of Theorem 6.4. We prove the other direction considering again the functions Ψ_t and $\tilde{\Psi}_t$ from the proof of Theorem 6.1. Actually, suppose that Π is a finitely connected hyperbolic domain in $\mathbb{C}$, $\tilde{C}_2(\Omega, \Pi) < \infty$, and (6.10) holds. Then

$$\frac{|\Psi_t''(z_0)|\,\mathrm{dist}(z_0, \partial\Omega)}{2\,\tilde{C}_2(\Omega, \Pi)} \leq \mathrm{dist}(e^{-t}, \partial\Pi) = e^{-t}.$$

If we let $t \to \infty$, we see that this contradicts (6.12), since the quotient $\mathrm{dist}(z_0, \partial\Omega)/R(z_0, \Omega)$ does not depend on t. A similar consideration in the case (6.11) reveals an analoguous contradiction. $\square$

The next theorem gives a further modification of Schwarz-Pick type inequalities for any pair of hyperbolic domains.

Theorem 6.5 ([21]). *Let Ω and Π be arbitrary hyperbolic domains in $\overline{\mathbb{C}}$. Then for any $f \in A(\Omega, \Pi)$, any $n \in \mathbb{N}$, and any $z \in \Omega \setminus \{\infty\}$ the estimate*

$$|f^{(n)}(z)| \leq \frac{4^n\,n!}{(\mathrm{dist}(f(z), \partial\Pi))^{n-1}} \left(\frac{R(f(z), \Pi)}{\mathrm{dist}(z, \partial\Omega)}\right)^n$$

is valid, if $f(z) \neq \infty$.

Proof. By the proof of Theorem 6.1, for $f \in A(\Omega, \Pi)$ and any fixed $z \in \Omega$ ($z \neq \infty$, $f(z) \neq \infty$) we have

$$(\text{dist}(z, \partial\Omega))^n |f^{(n)}(z)| = n! \, |a_n(g)|, \quad n \in \mathbb{N}, \tag{6.13}$$

where $g \in A(\Delta, \Pi)$ is defined as above and $a_0(g) = g(0) = f(z)$.

Now, let $w_0 = f(z)$ and let Φ_Π be a covering map from Δ to Π, such that Φ_Π is a locally univalent function on Δ and $\Phi'_\Pi(0) = R(w_0, \Pi)$. Moreover, for any simply connected domain $\Pi' \subset \Pi$, any branch

$$\Phi_\Pi^{-1}\big|_{\Pi'}$$

is a univalent function. We consider the branch

$$\Psi = \Phi_\Pi^{-1}\big|_{D(w_0, \rho)}, \quad D(w_0, \rho) = \{w \mid |w - w_0| < \rho = \text{dist}(w_0, \partial\Pi)\},$$

satisfying the condition $\Psi(w_0) = 0$. The Koebe one-quarter theorem implies that the function

$$h : \Delta \to \mathbb{C}, \; \zeta \mapsto h(\zeta) = \Psi(w_0 + \rho\zeta)\frac{R(w_0, \Pi)}{\rho}$$

attains all values in the disc $D_{1/4} = \{z \mid |z| < 1/4\}$. Hence,

$$D_a = \left\{z \mid |z| < a = \frac{\rho}{4\,R(w_0, \Pi)}\right\} \subset \Psi(D(w_0, \rho)). \tag{6.14}$$

Thus, Φ_Π is univalent on D_a.

As a consequence of the monodromy theorem, the function ω defined by

$$\omega : \Delta \to \Delta, \; \zeta \mapsto \omega(\zeta) = \Phi_\Pi^{-1} \circ g(\zeta), \quad \omega(0) = 0,$$

is a holomorphic self-map of Δ. Hence, $|\omega(\zeta)| \leq |\zeta|$ for all $\zeta \in \Delta$. This implies

$$g(D_a) \subset \Phi_\Pi(D_a) \subset D(w_0, \rho),$$

since $g(\zeta) = \Phi_\Pi(\omega(\zeta))$. Applying Cauchy's estimates for the Taylor coefficients of the function $g - w_0$ in the disc D_a, we get

$$a^n |a_n(g)| \leq \rho, \quad n \in \mathbb{N}. \tag{6.15}$$

From (6.13), (6.14), and (6.15) it follows that

$$(\text{dist}(z, \partial\Omega))^n |f^{(n)}(z)| \leq \frac{n! \, \rho}{a^n} \leq \frac{4^n \, n! \, (R(w_0, \Pi))^n}{\rho^{n-1}},$$

which is the desired inequality of Theorem 6.5. $\qquad\square$

Next, we shall examine some facts connected with Theorem 4.18.

Let Ω and Π be open sets on the Riemann sphere $\overline{\mathbb{C}}$ that are equipped with the Poincaré metric of curvature -4. According to Poincaré's generalization of Riemann's mapping theorem, this means that the boundaries of Ω and Π contain at least three points in $\overline{\mathbb{C}}$ and that the density of this metric in the unit disc $\Delta = \{z \mid |z| < 1\}$ is defined as

$$\lambda_\Delta(z) = \frac{1}{1 - |z|^2}, \quad z \in \Delta.$$

As usual, we consider the set $A(\Omega, \Pi)$ of functions $f : \Omega \to \Pi$, which are locally holomorphic or meromorphic and in general multivalued. Let $n \in \mathbb{N}$ and $Q_n(z, w, \Omega, \Pi)$ be defined as the smallest possible value such that the inequality

$$\frac{\left|f^{(n)}(z)\right|}{n!} \leq Q_n(z, w, \Omega, \Pi)\frac{(\lambda_\Omega(z))^n}{\lambda_\Pi(w)}$$

holds for all $f \in A(\Omega, \Pi)$, $f(z) = w$. For any pair $(z, w) \in \Omega \times \Pi$ and any $n \in \mathbb{N}$, normal family arguments give that there exists a function $f \in A(\Omega, \Pi)$ such that $f(z) = w$ and

$$\frac{\left|f^{(n)}(z)\right|}{n!} = Q_n(z, f(z), \Omega, \Pi)\frac{(\lambda_\Omega(z))^n}{\lambda_\Pi(f(z))}.$$

Moreover, it is easily seen that

$$C_n(\Omega, \Pi) = \sup\{Q_n(z, w, \Omega, \Pi) \mid (z, w) \in \Omega \times \Pi\}.$$

Here we shall need the function F defined in the proof of Theorem 4.18 by equation

$$F(p, q) := \begin{cases} p + q, & \text{if } p + 2q \geq 4, \\ 2 + p^2/(8 - 4q), & \text{if } p + 2q < 4, \end{cases} \tag{6.16}$$

for $p \geq 0$ and $q \geq 0$.

Theorem 6.6 ([27]). *Let Ω and Π be open sets on the Riemann sphere that are equipped with the Poincaré metric of curvature -4. Then for any $f \in A(\Omega, \Pi)$ the estimate*

$$|f''(z)| \leq F(p, q)\frac{(\lambda_\Omega(z))^2}{\lambda_\Pi(w)} \tag{6.17}$$

is valid for $w = f(z)$, $z \in \Omega' = \mathbb{C} \cap \Omega$, $w \in \Pi' = \mathbb{C} \cap \Pi$, $p = |\nabla(1/\lambda_\Omega(z))|$, and $q = |\nabla(1/\lambda_\Pi(w))|$. The inequality (6.17) is sharp for any $(z, w) \in \Omega' \times \Pi'$.

Proof. We fix $(z, w) \in \Omega' \times \Pi'$ and consider a function $f \in A(\Omega, \Pi)$, $f(z) = w$. Without loss of generality, we assume that Ω and Π equal their components that contain z and w, respectively. For this case the assertion follows from the proof of Theorem 4.18. $\qquad\square$

As in Chapter 5 we denote

$$\gamma(\Omega) = \sup_{z \in \Omega} |\nabla \left(1/\lambda_\Omega(z)\right)|, \quad \gamma(\Pi) = \sup_{w \in \Pi} |\nabla \left(1/\lambda_\Pi(w)\right)|$$

and we want to give some bounds for $C_2(\Omega, \Pi)$ dependent on the Euclidean geometry and on the conformal geometry of these sets.

From Chapter 3 and Theorem 6.6 we immediately have the following corollaries.

Corollary 6.7. *Let Ω and Π be open sets in $\mathbb{C}$ equipped with the usual Poincaré metric. Then $C_2(\Omega, \Pi)$ is finite if and only if $\overline{\mathbb{C}} \setminus \Omega$ and $\overline{\mathbb{C}} \setminus \Pi$ are uniformly perfect.*

Corollary 6.8. *Let Ω and Π be open sets in $\mathbb{C}$ equipped with the usual Poincaré metric. Then the following assertions are valid.*

(i) $C_2(\Omega, \Pi) = \frac{1}{2}(\gamma(\Omega) + \gamma(\Pi))$,

(ii) $2(M_0(\Omega) + M_0(\Pi)) \leq C_2(\Omega, \Pi) \leq 2\pi(M_0(\Omega) + M_0(\Pi)) + \Gamma(1/4)^4/\pi^2$,

(iii) $2(M(\Omega) + M(\Pi)) - 2 \leq C_2(\Omega, \Pi) \leq 2\pi(M(\Omega) + M(\Pi)) + \Gamma(1/4)^4/\pi^2$,

(iv) $C_2(\Omega, \Pi) \geq 2$, *where equality is attained if and only if all components of Ω and Π are convex.*

Together with the results of Chapter 4, where we proved that $C_n(\Omega, \Pi) = 2^{n-1}$ for convex domains Ω and Π, Corollary 6.8 (iv) delivers the proof of a weak form of a conjecture published in [16].

We want to formulate the remaining parts of a stronger conjecture as

Conjecture. *Let $n \in \mathbb{N} \setminus \{1, 2\}$ and let Ω and Π be open sets in $\mathbb{C}$ equipped with the usual Poincaré metric. Then $C_n(\Omega, \Pi) = 2^{n-1}$ implies that all components of Ω and Π are convex domains.*

Now, we turn to the cases where the point at infinity belongs to Ω or Π and we prepare these considerations with a lemma which may deserve some interest of its own.

Lemma 6.9 ([27]). *Let Ω be an open set in $\overline{\mathbb{C}}$ equipped with the usual Poincaré metric and let $\infty \in \Omega$. Then the equations*

(i) $\lim_{z \to \infty} |\nabla \left(1/\lambda_\Omega(z)\right)| = \infty$,

(ii) $\lim_{z \to \infty} |\nabla \log \lambda_\Omega(z)| \mathrm{dist}(z, \partial\Omega) = 2$,

are valid.

Proof. Without restriction of generality we assume Ω to be a hyperbolic domain. To prove (i) we consider a universal covering map f of Δ onto Ω such that $f(0) = \infty$. Let f have the expansion

$$z = f(\zeta) = \frac{a}{\zeta} + \varphi(\zeta), \quad \zeta \in \Delta,$$

where $a \neq 0$ and φ is a function meromorphic in Δ and holomorphic in a neighbourhood of the origin. Direct computations using

$$\lambda_\Omega(z) = \frac{1}{|f'(\zeta)|(1 - |\zeta|^2)}, \quad z = f(\zeta), \tag{6.18}$$

imply

$$|\nabla (1/\lambda_\Omega(z))| = \frac{1}{|\zeta|} \left| \frac{2a + \zeta^3 \varphi''(\zeta)}{a - \zeta^2 \varphi'(\zeta)} + 2|\zeta|^2 \right|. \tag{6.19}$$

If we take the limit $z \to \infty$ of the left side of equation (6.19), which is equivalent to taking the limit $\zeta \to 0$ of the right side, we see that (i) is valid.

To prove (ii) we multiply the left sides and the right sides of (6.18) and (6.19) with one another. If we multiply these products with $|z|$ and $|a/\zeta|$, we get

$$\lim_{z \to \infty} |\nabla \log \lambda_\Omega(z)||z| = \lim_{\zeta \to 0} |\nabla \log \lambda_\Omega(f(\zeta))| \left| \frac{a}{\zeta} \right| = 2. \tag{6.20}$$

Since $\partial\Omega$ is bounded,

$$\lim_{z \to \infty} \frac{\text{dist}(z, \partial\Omega)}{|z|} = 1. \tag{6.21}$$

The formulae (6.20) and (6.21) imply (ii). This completes the proof of Lemma 6.9. □

Lemma 6.9 (ii) gives rise to a sharp infinity version of the Osgood-Jørgensen theorem. For hyperbolic domains $\Omega \subset \mathbb{C}$ the fundamental inequality

$$|\nabla \log \lambda_\Omega(z)| \text{dist}(z, \partial\Omega) \leq 2, \quad z \in \Omega, \tag{6.22}$$

was discovered by Osgood (see [121], and compare also [111]). The proof uses certain results of Jørgensen (see [88]). One may observe that this proof in [121] and the related results in [88] do not use the fact that $\infty \notin \Omega$. Moreover, in [88] Jørgensen considers the case when $\infty \in \Omega$ as a basic one. This observation together with Lemma 6.9 (ii) leads us to the following theorem which proves (3.36).

Theorem 6.10 ([21]). *Let Ω be an open set in $\overline{\mathbb{C}}$ equipped with the usual Poincaré metric. For any $z \in \Omega \cap \mathbb{C}$ the inequality (6.22) is valid. The constant 2 in (6.22) cannot be replaced by a smaller one if $\infty \in \Omega$.*

6.3   Some examples

Let Π be a hyperbolic domain in the extended complex plane $\overline{\mathbb{C}}$.

For fixed $w \in \Pi \setminus \{\infty\}$ we consider the local Taylor expansions

$$f(z) = w + \sum_{n=1}^{\infty} a_n(w, f) z^n$$

valid in a neighbourhood of the origin. We will discuss some examples showing the behaviour of the quantities

$$A_n(w, \Pi) = \sup\{|a_n(w, f)| \mid f \in A(\Delta, \Pi), \ f(0) = w\}$$

regarded as a function of w (see [29]).

Example 1. We consider the doubly connected domain

$$\Pi = D_\infty := \{z \in \mathbb{C} \mid |z| > 1\}$$

and we prove that for $n \in \mathbb{N}$,

$$A_n(w, D_\infty) = F_n(-\log|w|), \qquad (6.23)$$

where

$$F_n(t) = -2t\, e^{-t} \sum_{k=0}^{n-1} \binom{n-1}{k} \frac{(-2t)^k}{(k+1)!}, \quad n \in \mathbb{N} \cup \{0\},$$

is Bateman's function of order n (compare [35] and [94]). The upper bound is attained if and only if there exist $\varphi_1, \varphi_2 \in \mathbb{R}$ such that

$$f(z) = e^{i\varphi_1} \Phi_{D_\infty, w}\left(e^{i\varphi_2} z\right).$$

Further,

$$\lim_{w \to \infty} \frac{A_n(w, D_\infty)(\mathrm{dist}(w, \partial D_\infty))^{n-1}}{(R(w, D_\infty))^n} = \frac{1}{n!}$$

and

$$\lim_{|w| \to 1} \frac{A_n(w, D_\infty)(\mathrm{dist}(w, \partial D_\infty))^{n-1}}{(R(w, D_\infty))^n} = 2^{1-n}.$$

To prove these assertions we use that, for $w = |w| e^{i\theta}$, $t =: \log|w| > 0$, the function $\Phi_{D_\infty, w}$ is defined by

$$\Phi_{D_\infty, w}(z) = e^{i\theta} \exp\left(t\, \frac{1 + e^{-i\theta} z}{1 - e^{-i\theta} z}\right)$$

and that for any $f \in A(\Delta, D_\infty)$, $f(0) = w$, there exists a uniquely defined function p holomorphic in Δ such that $\mathrm{Re}\, p(z) > 0$ for $z \in \Delta$, $p(0) = 1$, and

$$f(z) = e^{i\theta} \exp(t\, p(z)).$$

The first equation yields

$$R(w, D_\infty) = 2|w| \log|w| \qquad (6.24)$$

and the second by differentiation

$$f'(z) = t\, p'(z) f(z).$$

Using the Taylor coefficients of the expansion

$$p(z) = 1 + \sum_{k=1}^{\infty} p_k \, z^k,$$

we see that

$$a_1(w, f) = t \, w \, p_1, \quad n \, a_n(w, f) = t\Big(w \, n \, p_n + \sum_{k=1}^{n-1} a_k(w, f)(n-k)p_{n-k}\Big), \quad n \geq 2.$$

To prove (6.23) and the assertion on the extremal function we remember that the inequalities $|p_k| \leq 2$ are valid for $k = 1, \ldots, n$, with equality if and only if $p(z) = (1 + cz)/(1 - cz)$, $|c| = 1$, and the triangle inequality. Further we use the identity

$$n F_n(t) = -2t \sum_{k=0}^{n-1} (n-k) F_k(t)$$

which we were not able to find anywhere, but which may be easily proved by mathematical induction and the difference equation

$$(n-1)(F_n(t) - F_{n-1}(t)) + (n+1)(F_n(t) - F_{n+1}(t)) = 2t F_n(t)$$

(compare [35] and [94]).

Since

$$\lim_{t \to \infty} \frac{F_n(-t)}{(2t)^n e^t} = \frac{1}{n!},$$

the first assertion on the asymptotics of $A_n(w, D_\infty)$ is an immediate consequence of (6.23) and (6.24). The second one follows likewise.

This example shows that there are domains where the universal covering functions deliver the extrema for the moduli of all local Taylor coefficients.

Example 2. For the doubly connected domain

$$\Delta' = \{z \mid 0 < |z| < 1\}, \quad w = |w|e^{i\theta}, \quad t = -\log|w| > 0,$$

the universal covering function is given by

$$\Phi_{\Delta', w} = e^{i\theta} \exp\left(-t \frac{1 + e^{-i\theta} z}{1 - e^{-i\theta} z}\right)$$

and $f \in A(\Delta, \Delta')$, $f(0) = w$, may be written as

$$f(z) = e^{i\theta} \exp(-t \, p(z)),$$

where p is as in Example 1. It is easily seen by the same method as in Example 1 that

$$|a_n(w, \Phi_{D_0, w})| = |F_n(t)|.$$

This has been proved in [94].

Notwithstanding the fact that the determination of $A_n(w, \Delta')$ is very difficult the methods of Example 1 may be applied to get the right asymptotics near the origin. An analogous reasoning shows that for $f \in A(\Delta, \Delta')$, $f(0) = w$,

$$\lim_{|w| \to 0} \frac{|a_n(w, f)|(\mathrm{dist}(w, \partial\Delta'))^{n-1}}{(R(w, \Delta'))^n} = \frac{|p_1|^n}{2^n \, n!}$$

if we fix p in the above representation of f. Since $|p_1| \leq 2$ where equality is attained if f equals a universal covering function, we get

$$\lim_{|w| \to 0} \frac{A_n(w, \Delta')(\mathrm{dist}(w, \partial\Delta'))^{n-1}}{(R(w, \Delta'))^n} = \frac{1}{n!}.$$

In a similar way, we get

$$\lim_{|w| \to 1} \frac{|a_n(w, f)|(\mathrm{dist}(w, \partial\Delta'))^{n-1}}{(R(w, \Delta'))^n} = \frac{|p_n|}{2^n}$$

and conclude

$$\lim_{|w| \to 1} \frac{A_n(w, \Delta')(\mathrm{dist}(w, \partial\Delta'))^{n-1}}{(R(w, \Delta'))^n} = 2^{1-n}.$$

Example 3. Let $\delta > 1$ and the annulus A_δ be defined by

$$A_\delta = \{z \mid 1 < |z| < \delta\}.$$

It is known that the hyperbolic radius in this case has the form

$$R(w, A_\delta) = 2\,|w|\frac{\log \delta}{\pi} \sin \left(\pi \, \frac{\log |w|}{\log \delta} \right).$$

To get the asymptotics near the boundary we use the following representation of $f \in A(\Delta, A_\delta)$, $f(0) = w$. Let $\tau > 0$ and $t \in (-\pi/2, \pi/2)$ be chosen such that

$$w = \exp \left(\frac{\log \delta}{\pi} \left(t + \frac{\pi}{2} \right) \right) \exp \left(-i \, \frac{\log \delta \, \log \tau}{\pi} \right).$$

Then there exists a function p that is holomorphic and has positive real part in Δ, $p(0) = 1$, such that

$$f(z) = \exp \left(-i \, \frac{\log \delta}{\pi} \log \left(i \tau (\cos t \, p(z) + i \sin t) \right) \right).$$

Differentiation yields

$$(\cos t \, p(z) + i \sin t) f'(z) = -i \cos t \, \frac{\log \delta}{\pi} p'(z) f(z).$$

The consideration of the Taylor coefficients in this formula yields

$$a_1(w, f) = -i\, p_1\, w\, \frac{\log \delta}{\pi}\, \cos t$$

and

$$n\, a_n(w, f)e^{it} + \cos t \sum_{k=1}^{n-1} k\, a_k\, p_{n-k}$$

$$= -i\, \frac{\log \delta}{\pi}\, \cos t \left(n\, p_n\, w + \sum_{k=1}^{n-1} a_k(n-k)p_{n-k} \right).$$

By a reasoning analogous to that of the second example we get

$$\lim_{w \to \partial A_\delta} \frac{A_n(w, A_\delta)(\text{dist}(w, \partial A_\delta))^{n-1}}{(R(w, A_\delta))^n} = 2^{1-n}.$$

To get a global estimate in this case we use $A(\Delta, A_\delta) \subset A(\Delta, \Delta)$ for $w \in A_\delta$ and $|f(z)| < \delta$ for $z \in \Delta$. These facts yield immediately

$$A_n(w, A_\delta) \leq \min \left\{ F_n(-\log |w|), \frac{\delta^2 - |w|^2}{\delta} \right\}.$$

The difference between the first two examples and the third one lies in the fact that the quantity $R(w, \Omega)/\text{dist}(w\partial\Omega)$ is unbounded in the first two cases and bounded in the last one.

Chapter 7

Related results

7.1 Inequalities for schlicht functions

First, we will give an outline of the ideas and results that led to the conjectures of Chua. To our knowledge, E. Landau was the first who considered the possibility to follow G. Pick's program as indicated in the introduction for the higher derivatives of schlicht functions. He proved the following theorem (compare Landau [98], Gong [71]).

Theorem 7.1. *If the Bieberbach conjecture is valid for $n \geq 2$, then for these n, any $z_0 \in \Delta$, and any function f holomorphic and injective on Δ, the inequality*

$$\frac{|f^{(n)}(z_0)|}{n!} \leq |f'(0)| \frac{n + |z_0|}{(1 - |z_0|)^{n+2}} \tag{7.1}$$

is valid.

It is clear that, on the other hand, the validity of the Bieberbach conjecture follows from (7.1). The proof uses automorphisms of the unit disc in the same way as the formula of Szász mentioned above. Since the same method was used to prove a similar formula of Jakubowski that originated in the first half of the last century, we give a unified proof for these two theorems after the presentation of Jakubowski's Theorem (see [85]).

Theorem 7.2. *If the Bieberbach conjecture is valid for $n \geq 2$, then for these n, any $z_0 \in \Delta$, and any function f holomorphic and injective on Δ, the inequality*

$$\frac{|f^{(n)}(z_0)|}{n!} \leq |f'(z_0)| \frac{(n + |z_0|)(1 + |z_0|)^{n-2}}{(1 - |z_0|^2)^{n-1}} \tag{7.2}$$

is valid.

Proof of Theorems 7.1 *and* 7.2. Let f be injective on Δ. The function

$$g(\zeta) = \frac{f\left(\frac{\zeta + z_0}{1 - \overline{z_0}\zeta}\right)}{(1 - |z_0|^2)f'(z_0)}$$

is injective on Δ and $g'(0) = 1$. Using Szasz's formula and de Branges' theorem for

$$g(\zeta) = g(0) + z + \sum_{k=2}^{\infty} a_k \zeta^k$$

we get immediately, as in Chapter 4,

$$\frac{|f^{(n)}(z_0)|(1 - |z_0|^2)^n}{(1 - |z_0|^2)|f'(z_0)|\,n!} \leq \sum_{k=1}^{n} \binom{n-1}{n-k} |z_0|^{n-k} k = (n + |z_0|)(1 + |z_0|)^{n-2}.$$

Apparently, this is Jakubowski's theorem.

Since Landau was interested in an upper bound dependent on $|z_0|$ only, he used in his proof the distortion theorem

$$\frac{|f'(z_0)|}{|f'(0)|} \leq \frac{1 + |z_0|}{(1 - |z_0|)^3}$$

for functions injective on Δ. Combining this with Jakubowski's bound delivers Landau's theorem.   □

The case of equality in Jakubowski's results has been discussed by Yamashita in [169] in detail.

Further, Jakubowski recognized that by the same method, using Löwner's theorem for the MacLaurin coefficients of functions f convex and injective on Δ, one gets the bound

$$\frac{|f^{(n)}(z_0)|}{n!} \leq |f'(z_0)|\frac{(1 + |z_0|)^{n-1}}{(1 - |z_0|^2)^{n-1}}.$$

Here, again the case of equality is due to Yamashita (see [169]).

A natural simplifying of these bounds is achieved, if one replaces in the numerator $|z_0|$ by 1. This results in the factors $(n + 1)2^{n-2}$ in the case of injective functions and 2^{n-1} in the case of convex functions. These constants play an important part in the present book, Chapter 5.

Possibly, the resulting formula, the identity

$$R(z_0, \Delta) = 1 - |z_0|^2$$

and naturally de Branges' theorem motivated Chua in 1996 to consider functions f injective or convex on simply connected proper subdomains Ω of $\mathbb{C}$ (see [56]).

He proceeded as we did in Chapter 4 with the exception that for him there was no need to consider subordination. Therefore he arrived at a formula analogous to our formula (4.5). The first difference is that in his formula there appear the coefficients of schlicht or convex functions and not those of subordinates to them. The second difference is that he could use the identity

$$R(f(z_0), f(\Omega)) = |f'(z_0)| \, R(z_0, \Omega).$$

Concerning a formula equivalent to the formula of Szász that we used, he found a formula proved by Todorov in [163]. Naturally, he had to consider the coefficients of powers of inverses of schlicht functions, as we do. He knew the proofs of Löwner (see [110] and of Schober (see [147]) for the coefficients of the inverses themselves, but he was not aware of the fact that the extremality of the Koebe function for powers, where the exponent is a natural number, is a simple consequence of Löwner's theorem. Therefore, he proved this fact using Baernstein's integral mean theorem (see [32]) and the Schur-Jabotinski theorem (see [81], Thm 1.9.a). By these theorems, he could prove that for f injective on Ω, $z_0 \in \Omega$, the inequality

$$\frac{|f^{(n)}(z_0)|}{n!} \le |f'(z_0)| \, \frac{4^{n-1}}{(R(z_0, \Omega))^{n-1}}, \quad n \ge 2,$$

is valid , whereas for f convex and injective he got

$$\frac{|f^{(n)}(z_0)|}{n!} \le |f'(z_0)| \, \frac{\binom{2n-1}{n}}{(R(z_0, \Omega))^{n-1}}, \quad n \ge 2.$$

Further, Chua recognized that, for the consideration of similar problems for functions f injective or convex on convex proper subdomains of $\mathbb{C}$ that led him to his conjectures, one needs bounds for the inverses of convex functions. Using a theorem due to Trimble (see [164]) he suceeded in proving the conjectures up to $n = 4$. The results of Li (see [102] and [103]), who used theorems of Libera and Zlotkiewicz (see [104]) on the coefficients of inverses of convex functions, imply that these conjectures are valid for $n \le 8$.

As we have seen in Chapter 4, the computation of punishing factors is possible, if one has a good knowledge of the inverse coefficients of certain functions on one hand and good results on the coefficients of subordinated functions on the other hand.

If one reduces the interest to the conformal maps themselves, one can replace the subordination results by theorems on the bounds of coefficients of certain injective functions. We want to cite two examples where the proofs of the details are the same as in Chapter 4 and Chapter 5. In addition we need the interplay between a geometric property, the accessibility of order β and an analytic property, the close-to-convexity of order β.

Definition 7.3. A domain Π is called (angularly) accessible of order β, $\beta \in [0, 1]$, if it is the complement of a union of rays that are pairwise disjoint except that the origin of one ray may lie on another one of the rays, and such that every ray is the bisector of a sector of angle $(1 - \beta)\pi$ that wholly lies in the complement of Π.

We use the following characterization of domains accessible of order β.

Theorem 7.4 (see for instance [127] and [93] and compare [41]). *Let f be injective and holomorphic in the unit disc and $f(\Delta)$ accessible of order β. Then there exist $\alpha \in [0, 2\pi]$, and functions g and p holomorphic in Δ such that the following conditions hold:*

(i) *$g(\Delta)$ is convex,*

(ii) *$\operatorname{Re}(e^{i\alpha}p(z)) > 0$ for $z \in \Delta$ and $p(0) = 1$,*

(iii) *$f'(z) = (p(z))^\beta g'(z)$ for $z \in \Delta$.*

Functions satisfying the proprties (i)–(iii) are called (nonnormalized) functions close-to-convex of order β and we owe to Ahoronov and Friedland the proof (see [2] and compare [146]) that for such functions the following inequalities hold.

Theorem 7.5. *Let f be close-to-convex of order β. Then*

$$\frac{|f^{(n)}(0)|}{|f'(0)|} \le \frac{1}{2(\beta + 1)} \frac{d^n}{(dz)^n} \left. \left(\frac{1+z}{1-z} \right)^{1+\beta} \right|_{z=0}, \quad n \in \mathbb{N}.$$

These theorems and the considerations and computations of Chapter 5 immediately imply the following theorems.

Theorem 7.6. *Let f be holomorphic and injective on a simply connected proper subdomain of $\mathbb{C}$ and such that $f(\Delta)$ is linearly accessible of order β, $\beta \in [0, 1]$. Then for $z_0 \in \Omega$ and $n \in \mathbb{N}$ the inequalities*

$$\frac{|f^{(n)}(z_0)|}{|f'(z_0)|} \le \frac{1}{(R(z_0, \Omega))^{n-1}} \frac{4^{n-1}}{\beta + 1} \binom{\frac{\beta+1}{2} + n - 1}{n}$$

are valid.

Theorem 7.7. *Let f be holomorphic and injective on a simply connected proper subdomain Ω of $\overline{\mathbb{C}}$, $\infty \in \Omega$, and such that $f(\Delta)$ is linearly accessible of order β, $\beta \in [0, 1]$. Then for $z_0 \in \Omega \setminus \{\infty\}$ and $n \in \mathbb{N}$ and $p(z_0)$ defined as above, the inequalities*

$$\frac{|f^{(n)}(z_0)|}{|f'(z_0)|} \le \frac{1}{(R(z_0, \Omega))^{n-1}} \sum_{k=1}^{n} \frac{4^k}{2(\beta + 1)} \binom{n-1}{n-k} \binom{\frac{\beta+1}{2}}{k} \left(p(z_0) + \frac{1}{p(z_0)} + 2 \right)^{n-k}$$

are valid.

7.2 Derivatives of α-invariant functions

Let Ω be a simply connected domain in $\overline{\mathbb{C}}$ with the density λ_Ω of the Poincaré metric. Our aim is to find bounds for the function

$$M_n(a, U_\alpha(\Omega)) = \frac{1}{n!} \max\{|f^{(n)}(a)| : f \in U_\alpha(\Omega)\}, \quad a \in \Omega,$$

where α is a parameter and $U_\alpha(\Omega)$ is a family of holomorphic functions with a given property. For instance, $U_\alpha(\Omega)$ is the unit ball $\|f\| \leq 1$ in one of the usual spaces of holomorphic functions.

To find M_n we will use the functions (see Chapter 3)

$$\mu_k = \mu_k(a, \Omega) = \frac{1}{(k-1)!} \lambda_\Omega^{-k}(a) \frac{\partial^k \log \lambda_\Omega^2(a)}{\partial a^k}, \quad k = 1, \ldots, n.$$

We start with a simple remark. For all conformal automorphisms $\varphi : \Omega \to \Omega$, the condition $f \in A(\Omega, \Pi)$ implies that the functions $f \circ \varphi$ are members of $A(\Omega, \Pi)$, too. With this observation in mind, we shall consider the following generalization of linear invariant families by Ch.Pommerenke [126] (compare also [8]).

The first definition deals with functions holomorphic in the unit disc Δ.

Definition 7.8. Let $\alpha = \text{const} \geq 0$, and let U_α be a set of functions g such that:

 (i) g is holomorphic in Δ,

 (ii) for all conformal automorphisms $\varphi : \Delta \to \Delta$,

$$g \in U_\alpha \Rightarrow g_{\alpha, \varphi} \in U_\alpha,$$

   where $g_{\alpha,\varphi}(\zeta) := g(\varphi(\zeta))\varphi'^\alpha(\zeta)$,

 (iii) $g(\zeta)$ are uniformly bounded in the interior of Δ and $(0, \ldots, 0)$ is an interior point of the coefficient region

$$\{(a_0, \ldots, a_n) : g(\zeta) = a_0 + a_1\zeta + \cdots \in U_\alpha\}.$$

The following definition describes α- invariant families of functions holomorphic in a simply connected domain Ω.

Definition 7.9. Given U_α, by $U_\alpha(\Omega)$ we denote the set of all functions f such that f is holomorphic in Ω and $f_{\alpha, \Phi} \in U_\alpha$ for all conformal mappings Φ of Δ onto Ω, where $f_{\alpha, \Phi}(\zeta) = f(\Phi(\zeta))\Phi'^\alpha(\zeta)$. If $\infty \in \Omega$, then we suppose that $2\alpha \in \mathbb{N} \cup \{0\}$.

Remark 7.10. The set $A(\Omega, \Pi)$ is a 0-invariant family.

Let $P_w(\zeta)$ be the polynomial $1 + w_1\zeta + \cdots + w_n\zeta^n$. For $g(\zeta) = a_0 + a_1\zeta + \cdots \in U_\alpha$, denote $g_n(\zeta) = a_n + a_{n-1}\zeta + \cdots + a_0\zeta^n$. We will consider the Hadamard product

$$(g_n * P_w)(\zeta) = a_n + a_{n-1}w_1\zeta + \cdots + a_0w_n\zeta^n.$$

Theorem 7.11 ([14]). *If $a \in \Omega$ and $f(z) \in U_\alpha(\Omega)$, then*

$$M_n(a, U_\alpha(\Omega)) = \lambda_\Omega^{n+\alpha}(a) \max_{g \in U_\alpha} \max_{|\zeta|=1} |(g_n * P_w)(\zeta)|, \tag{7.3}$$

where $w = (\tau_{n,n-1}(\alpha), \tau_{n,n-2}(\alpha), \ldots, \tau_{n,0}(\alpha))$, $\tau_{n,k}(\alpha)$ are defined by the following recurrent formulas:

$$\tau_{k,k}(\alpha) = 1, \ 0 \le k \le n, \quad \tau_{k,0}(\alpha) = \frac{\alpha}{k} \sum_{s=0}^{k-1} \mu_{k-s} \tau_{s,0}(\alpha), \ 1 \le k \le n, \tag{7.4}$$

$$\tau_{m,k}(\alpha) = \sum_{s=1}^{m-k+1} \frac{1}{s} \tau_{s-1,0}(1) \tau_{m-s,k-1}(\alpha), \ 2 \le k \le m \le n. \tag{7.5}$$

Proof. Let ψ be the conformal mapping of Ω to Δ, $\psi(a) = 0$, $\psi'(a) = \lambda_\Omega(a) > 0$. As in Chapter 3, we have

$$\frac{\partial \log \lambda_\Omega^2(z)}{\partial z} = \frac{\psi''(z)}{\psi'(z)} - \frac{\overline{2\psi(z)}}{1 - \psi(z)\overline{\psi(z)}} \psi'(z).$$

Hence

$$\lambda_\Omega^k(a) \mu_k(a, \Omega) = \frac{1}{(k-1)!} \frac{\partial^k \log \lambda_\Omega^2(a)}{\partial a^k} = \frac{1}{(k-1)!} \left(\frac{\psi''(z)}{\psi'(z)} \right)^{(k-1)} \bigg|_{z=a} \tag{7.6}$$

and

$$\frac{\psi''(z)}{\psi'(z)} = \sum_{k=0}^{\infty} \lambda_\Omega^{k+1}(a) \mu_{k+1}(a, \Omega)(z-a)^k \tag{7.7}$$

in some neighbourhood of a.

Consider the function

$$g(\zeta) = f(\Phi(\zeta)) \Phi'^{\alpha}(\zeta) = a_0 + a_1 \zeta + \ldots, \qquad \zeta \in \Delta, \tag{7.8}$$

where $z = \Phi(\zeta)$ is the inverse of the function $\zeta = \psi(z)$. It follows from (7.8) that

$$\frac{1}{m!} f^{(m)}(a) = \frac{1}{m!} \sum_{k=0}^{\infty} a_k (\psi^k(z) \psi'^{\alpha}(z))^{(m)} \bigg|_{z=a}$$

$$= \sum_{k=0}^{m} a_k t_{m,k}(\alpha)$$

with $m! \, t_{m,k}(\alpha) = (\psi^k(z) \psi'^{\alpha}(z))^{(m)}|_{z=a}$. We have to prove that

$$t_{m,k}(\alpha) / \lambda_\Omega^{m+\alpha}(a) = \tau_{m,k}(\alpha)$$

satisfy (7.4) and (7.5) for $m = 1, 2, \ldots, n$.

From

$$\psi(z) = \lambda_\Omega(a)(z-a)[1 + O(z-a)],$$

we have directly $t_{k,k}(\alpha) = \lambda_\Omega^{k+\alpha}(a)$ in accordance with $\tau_{k,k}(\alpha) = 1$. Using (7.6) and the Leibniz formula, we have

$$t_{k,0}(\alpha) = \frac{1}{k!}(\psi'^\alpha(z))^{(k)}\Big|_{z=a} = \frac{\alpha}{k!}\left(\psi'^\alpha(z)\frac{\psi''(z)}{\psi'(z)}\right)^{(k-1)}\Big|_{z=a}$$

$$= \frac{\alpha}{k!}\sum_{s=0}^{k-1}\binom{k-1}{s}(\psi'^\alpha(z))^{(s)}\Big|_{z=a} \cdot \left(\frac{\psi''(z)}{\psi'(z)}\right)^{(k-1-s)}\Big|_{z=a}$$

$$= \frac{\alpha}{k}\sum_{s=0}^{k-1}\lambda_\Omega^{k-s}(a)\mu_{k-s}t_{s,0}(\alpha).$$

We also have

$$t_{m,k}(\alpha) = \frac{1}{m!}\sum_{s=0}^{m}\binom{m}{s}(\psi(z))^{(s)}\Big|_{z=a}(\psi^{k-1}(z)\psi'^\alpha(z))^{(m-s)}\Big|_{z=a}$$

$$= \sum_{s=1}^{m-k+1}\frac{1}{s!}t_{m-s,k-1}(\alpha)(\psi'(z))^{(s-1)}\Big|_{z=a} = \sum_{s=1}^{m-k+1}\frac{1}{s}t_{m-s,k-1}(\alpha)t_{s-1,0}(1).$$

Mathematical induction gives (7.4) and (7.5) for $\tau_{n,k}(\alpha) = t_{n,k}(\alpha)/\lambda_\Omega^{n+\alpha}(a)$. Thus,

$$\frac{1}{n!}f^{(n)}(a) = \lambda_\Omega^{n+\alpha}(a)(g_n * P_w)(1)$$

for $w = (\tau_{n,n-1}(\alpha), \tau_{n,n-2}(\alpha), \ldots, \tau_{n,0}(\alpha))$, $g(\zeta) = f(\Phi(\zeta))\Phi'^\alpha(\zeta)$. Taking into account the property

$$g(\zeta) \in U_\alpha \Rightarrow g(e^{-i\theta}\zeta) \in U_\alpha,$$

we obtain

$$M_n(a, U_\alpha(\Omega)) = \frac{1}{n!}\max\{|f^{(n)}(a)| : f \in U_\alpha(\Omega)\}$$

$$= \lambda_\Omega^{n+\alpha}(a)\max_{g \in U_\alpha}|(g_n * P_w)(1)| = \lambda_\Omega^{n+\alpha}(a)\max_{g \in U_\alpha}|(g_n * P_w)(e^{i\theta})|$$

$$= \lambda_\Omega^{n+\alpha}(a)\max_{g \in U_\alpha}\max_{|\zeta|=1}|(g_n * P_w)(\zeta)|.$$

This completes the proof of Theorem 7.11. $\qquad\qquad\qquad\qquad\qquad\square$

Using the well-known properties of $\max\{|F(\zeta)| : |\zeta| = r\}$ $(0 < r < \infty)$ for the analytic function $F(\zeta) = (g_n * P_w)(\zeta)$ (see [59]), we obtain the following corollary .

Corollary 7.12. *Let* $w = (\tau_{n,n-1}(\alpha), \tau_{n,n-2}(\alpha), \ldots, \tau_{n,0}(\alpha))$. *The function*

$$K_n(w, U_\alpha) = M_n(a, U_\alpha(\Omega))/\lambda_\Omega^{n+\alpha}(a)$$

has the following properties:

(i) $K_n(w, U_\alpha)$ *depends on* a *and* Ω *via* w *only,*

(ii) $K_n(w, U_\alpha)$ *is a convex function from* $\mathbb{C}^n(= \mathbb{R}^{2n})$ *to* $(0, \infty)$,

(iii) *there exist two positive constants* $c(n, \alpha)$ *and* $C(n, \alpha)$ *such that*

$$c(n, \alpha) \leq K_n(w, U_\alpha)/\sqrt{1 + \|w\|^2} \leq C(n, \alpha), \quad w \in \mathbb{C}^n,$$

(iv) $K(w_1, \ldots, w_n, U_\alpha) = K(e^{i\theta} w_1, \ldots, e^{in\theta} w_n, U_\alpha)$ *for any* $\theta \in [0, 2\pi]$,

(v) *for* $w \in \mathbb{C}^n$, *the function* $u(r) = K_n(rw_1, r^2 w_2, \ldots, r^n w_n, U_\alpha)$ *is nondecreasing on* $0 \leq r \leq \infty$ *and* $\log u(r)$ *is convex with respect to* $\log r$.

In the case $0 \leq k \leq n - 1$, $\tau_{n,k}(\alpha)$ depends on n, k, α, a and Ω. For example, if $\Omega = \Delta$ and $a \in \Delta$, then

$$\tau_{n,k}(\alpha) = \left(\begin{array}{c} n + 2\alpha - 1 \\ n - k \end{array} \right) \bar{a}^{n-k}.$$

Note that $\tau_{n,k}(\alpha)$ does not depend on $a \in \Omega$ if and only if Ω is a half-plane.

In the general case, for domains $\Omega \subset \mathbb{C}$ we proved that $\tau_{n,k}(\alpha)$ are bounded. Namely, by Theorem 3.27 we have the following assertion.

If $\Omega \subset \mathbb{C}$, *then*

$$\sup_{\Omega} \sup_{a \in \Omega} |\tau_{n,k}(\alpha)| = \sum_{s=0}^{n-k} \left(\begin{array}{c} 2n + 3\alpha - 1 \\ s \end{array} \right) \left(\begin{array}{c} n - k - s + \alpha - 2 \\ n - k - s \end{array} \right). \tag{7.9}$$

From Theorem 7.11 and Theorem 3.27 one immediately obtains the following corollary.

Corollary 7.13. *There exists an increasing convex function* $K_1(., U_\alpha) : [0, \infty) \to (0, \infty)$ *such that*

$$M_1(a, U_\alpha(\Omega)) = \lambda_\Omega^{1+\alpha}(a) K_1(t, U_\alpha), \qquad t = \alpha |\nabla \lambda_\Omega^{-1}(a)|. \tag{7.10}$$

If $\Omega \subset \mathbb{C}$, *then*

$$\sup_{\Omega} \sup_{a \in \Omega} \frac{M_1(a, U_\alpha(\Omega))}{\lambda_\Omega^{1+\alpha}(a)} = K_1(4\alpha, U_\alpha). \tag{7.11}$$

Consider now Hardy, Bergman and Bloch spaces (see for instance [59], [7]). We use the following notation:

$$\|f\|_p = \sup_{0<r<1} \left(\int_{|\psi(z)|=r} |f(z)|^p |dz| \right)^{1/p} \quad \text{for} \quad f \in E_p(\Omega),\, p > 0,$$

where ψ is the conformal mapping of Ω onto Δ, $\psi(a) = 0$, $\psi'(a) > 0$;

$$\|f\|_{p,\gamma} = \left(\frac{\gamma-1}{\pi} \iint_\Omega |f(z)|^p \lambda_\Omega^{\gamma-2}(z)\, dx\, dy \right)^{1/p}, \quad p > 0,\, \gamma > 1,$$

for $f \in A_{p,\gamma}(\Omega)$;

$$\|f\|_p^* = \sup_{z\in\Omega} |f(z)/\lambda_\Omega^{1/p}(z)| \quad \text{for} \quad f \in B_p^*(\Omega),\, p > 0.$$

Theorem 7.14 ([14]). *If $t = |\nabla \lambda_\Omega^{-1}(a)|$, then for any $a \in \Omega$,*

$$|f'(a)| \leq \lambda_\Omega^{1+1/p}(a) \left(1 + \frac{t}{p} \right) \|f\|_p, \quad p \geq 1, \tag{7.12}$$

$$|f'(a)| \leq \lambda_\Omega^{1+1/p}(a) \left((2-p)^{1/p-1/2} 2^{1-1/p} p^{-1/2} + \frac{t}{p} \right) \|f\|_p, \quad 0 < p \leq 1, \tag{7.13}$$

$$|f'(a)| \leq \lambda_\Omega^{1+1/p}(a) \left((2+p)^{1/p+1/2} 2^{-1/p} p^{-1/2} + \frac{t}{p} \right) \|f\|_p^*, \quad p > 0, \tag{7.14}$$

$$|f'(a)| \leq \lambda_\Omega^{1+1/(\gamma p)}(a) \left(\frac{2\Gamma(\gamma+1/2)}{\sqrt{\pi}\Gamma(\gamma)} + \frac{t}{\gamma p} \right) \|f\|_{p,\gamma}, \quad p \geq 1,\, \gamma > 1, \tag{7.15}$$

where Γ denotes Euler's gamma function.
The estimates (7.12), (7.13), (7.14), (7.15) are asymptotically sharp as $t \to +\infty$, (7.12), (7.13) and (7.14) are also sharp at $t = 0$.

Note that if $U_{1/p}$ is the unit ball in the Hardy space H_p, then the function $K_1(t, U_{1/p})$ is known by [114] in the cases $p \geq 1+t/2$ and $1 \leq p \leq \max\{2, 1+t/2\}$. For Bloch space $B_p^*(\Omega)$ we obtain from Corollary 7.13 and [165],

$$\max_{\|f\|_p^* \leq 1} |f'(z)| = (2+p)^{1/p+1/2} 2^{-1/p} p^{-1/2} \lambda_\Omega^{1+1/p}(a) k_p(|\text{grad}\,\lambda_\Omega^{-1}(a)|), \tag{7.16}$$

where $k_p(t) = \max_{0 \leq x \leq \sqrt{1+2/p}} [1 + tx/p - (1+2/p)x^2](1-x^2)^{1/p}$.

Proof of Theorem 7.14. From Theorem 7.11 we obtain

$$M_1(a, U_\alpha(\Omega)) = \sup_{g\in U_\alpha} \left(|a_1| + |\tau_{1,0}(\alpha)||a_0| \right) \lambda_\Omega^{1+\alpha}(a),$$

where $a_0 = g(0)$, $a_1 = g'(0)$, $|\tau_{1,0}(\alpha)| = \alpha |\nabla \lambda_\Omega^{-1}(a)|$. If $\alpha = 1/p$, $p \geq 1$ and $g \in H_p$, then $|a_0| \leq \|g\|_p$ and $|a_1| \leq \|g\|_p$ (see [59]). Taking into account Corollary 7.13, we have (7.12).

If $\alpha = 1/p$, $p > 0$ and $g \in B_p^*(\Delta)$, then the region $\{(|a_0|, |a_1|) : \|g\|_p^* \leq 1\}$ is given in [165], and we get (7.16) and (7.14).

Consider the case $g \in A_{p,\gamma}(\Delta)$, $p \geq 1$. By Hardy's theorem,

$$2\pi |g(0)|^p \leq \int_0^{2\pi} |g(re^{i\theta})|^p d\theta, \qquad 0 < r < 1.$$

This yields

$$\|g\|_{p,\gamma} \geq |a_0| \left(\frac{\gamma - 1}{\pi} \int_0^1 2r(1 - r^2)^{\gamma - 2} dr \right)^{1/p} = |a_0|.$$

We also have

$$\|g\|_{p,\gamma} \geq \|g\|_{1,\gamma} \geq 2|a_1|(\gamma - 1) \int_0^1 r^2 (1 - r^2)^{\gamma - 2} dr$$
$$= |a_1| \sqrt{\pi} \Gamma(\gamma) / (2\Gamma(\gamma + 1/2)).$$

Hence, (7.15) holds.

If $g(\zeta) \in H_p$ $(0 < p < 1)$, then $|a_0| \leq \|g\|_p$ (see [59]). Using the estimates from [157], we have $|a_1| \leq (2 - p)^{1/p - 1/2} 2^{1 - 1/p} p^{-1/2}$. Thus, (7.13) holds. $\square$

Theorem 7.15 ([14]). *If $f \in A_{2,\gamma}(\Omega)$, then for any $a \in \Omega$ and any $\gamma \geq 1$,*

$$\frac{1}{n!} \max_{\|f\|_{2,\gamma} \leq 1} |f^{(n)}(a)| = \lambda_\Omega^{n + \frac{1}{2\gamma}}(a) \left(\sum_{k=0}^{n} M \binom{k + \gamma - 1}{k} \left| \tau_{n,k} \left(\frac{1}{2\gamma} \right) \right|^2 \right)^{1/2}. \quad (7.17)$$

Proof. By direct calculation, we obtain

$$\|g\|_{2,\gamma} = \left(\sum_{k=0}^{\infty} \binom{k + \gamma - 1}{k}^{-1} |a_k|^2 \right)^{1/2}.$$

By the Cauchy-Schwarz inequality

$$\left| \sum_{k=0}^{n} a_k \tau_{n,k}(\alpha) \right| \leq \|g\|_{2,\gamma} \left(\sum_{k=0}^{n} \binom{k + \gamma - 1}{k} |\tau_{n,k}(\alpha)|^2 \right)^{1/2}.$$

Equality holds for $g(\zeta) = a_0 + a_1 \zeta + \cdots + a_n \zeta^n$ if and only if

$$\binom{k + \gamma - 1}{k} |\tau_{n,k}(\alpha)| = a_k \, const, \qquad k = 0, 1, \ldots, n,$$

which completes the proof of (7.17). $\square$

Theorem 7.16 ([14]). *If $f \in E_p(\Omega)$, then for all $a \in \Omega$,*

$$c(n,p) \leq \frac{1}{n!} \max_{\|f\|_p \leq 1} \frac{|f^{(n)}(a)|}{\lambda_\Omega^{n+1/p}(a)(\sum_{k=1}^n |\tau_{n,k}(\frac{1}{p})|^2)^{1/2}} \leq C(n,p)$$

with the constants $c(n,p)$ and $C(n,p)$ such that

$$c(n,p) = 1, \quad C(n,p) = const\ n^{1/p-1/2} \ for \ 0 < p \leq 2$$

and

$$c(n,p) = n^{1/2-1/p}, \quad C(n,p) = 1 \ for \ 2 \leq p \leq \infty.$$

Remark 7.17. For $p = 2$, Theorem 7.15 and Theorem 7.16 are well known (see for instance [114]).

Proof of Theorem 7.16. It is obvious that $c(n,2) = C(n,2) = 1$. Hence, $c(n,p) \geq 1$ for $0 < p \leq 2$ and $C(n,p) \leq 1$ for $2 \leq p \leq +\infty$, because $H_p \subset H_x$ for $p > x$.

Let $p > 2$. Hence, $\|g\|_2 \leq \|g\|_p$. By the Hausdorff-Young inequality, we have

$$\|g_n\|_p \leq \left(\sum_{k=0}^n |a_k|^q\right)^{1/q} \qquad \left(\frac{1}{p} + \frac{1}{q} = 1\right).$$

Consequently, by the Hölder inequality,

$$\|g_n\|_p \leq (n+1)^{1/2-1/p} \left(\sum_{k=0}^n |a_k|^2\right)^{1/2}.$$

Thus, $c(n,p) \geq n^{1/2-1/p}$ for $p > 2$.

To complete the proof of Theorem 7.16, we have to prove that $C(n,p) = O(n^{1/p-1/2})$ for $0 < p < 2$.

Let $0 < p \leq 1$. If $g(\zeta) = a_0 + a_1\zeta + \cdots \in H_p$ and $\|g\|_p \leq 1$, then by the Hardy-Littlewood theorem (see Theorem 6.5 in [59]) $a_n = o(n^{1/p-1})$. Consequently,

$$C(n,p) = \max_{\|g\|_p \leq 1} (\sum_{k=0}^n |a_k|^2)^{1/2} = O(n^{1/p-1/2}).$$

Let $1 < p < 2$. If $\|g\|_p \leq 1$, then by the Hausdorff-Young inequality,

$$\left(\sum_{k=0}^\infty |a_k|^q\right)^{1/q} \leq \|g\|_p \qquad \left(\frac{1}{p} + \frac{1}{q} = 1\right).$$

Using the Hölder inequality, we obtain $C(n,p) = O(n^{1/p-1/2})$. $\square$

7.3   A characterization of convex domains

Everybody who reads Chapter 4, Section 5 very carefully will recognize that, for the proof of the 2^{n-1} conjecture, we only need one of the Marx-Strohhäcker theorems. This inequality characterises the functions of the closed convex hull of the convex functions, but not the convex functions themselves. On the other hand, we know from Chapter 4, Section 6 that convex pairs are the only ones for which $C_2(\Omega, \Pi) = 2$. At first glance, this seems to be a contradiction. Thinking of this problem, we found that an invariant formulation of the mentioned Marx-Strohhäcker inequality in fact characterises convexity. The proof of this fact forms the content of this section and it seems to explain the geometrical background of Theorem 4.15.

Let g_0 be a function holomorphic and univalent in the unit disc Δ. We suppose that $g_0(0) = 0$, $g_0'(0) = 1$ and $\Pi := g_0(\Delta)$ is a convex domain. To characterize Π we shall use the fact that for such a function g_0 one of the well-known Marx-Strohhäcker inequalities, namely

$$\mathrm{Re}\left(\frac{g_0(z)}{z}\right) > \frac{1}{2}, \quad z \in \Delta, \tag{7.18}$$

holds (see [115] and [155]). For unified proofs of this and many related inequalities one should consult [118].

The formula (7.18) is equivalent to the existence of a bounded holomorphic function $\omega : \Delta \to \overline{\Delta}$ such that

$$\frac{g_0(z)}{z} = \frac{1}{1 + z\omega(z)}, \quad z \in \Delta. \tag{7.19}$$

We will consider the following question. Let h be a function holomorphic on Δ. Suppose that $h'(\zeta) \neq 0$ for any $\zeta \in \Delta$ and that $h(\Delta)$ has the Marx-Strohhäcker property for any point $z_0 = h(t) \in h(\Delta)$, i.e., the function g_0 defined by the Koebe transform

$$g_0(w) = \frac{h\left(\frac{w+t}{1+\bar{t}w}\right) - h(t)}{h'(t)(1 - |t|^2)}$$

satisfies inequality (7.18) for any $t \in \Delta$. This is equivalent to the inequality

$$\mathrm{Re}\left(\frac{h(z) - h(t)}{h'(t)} \frac{1 - \bar{t}z}{z - t}\right) > \frac{1 - |t|^2}{2}, \quad z \in \Delta, t \in \Delta. \tag{7.20}$$

What can be said about $\Omega = h(\Delta)$? We find that $h(\Delta)$ is a convex domain, so that an assertion inverse to the Marx-Strohhäcker theorem is valid.

Theorem 7.18 ([24]). *Let h be a function holomorphic in Δ, such that $h'(\zeta) \neq 0$, $\zeta \in \Delta$, and such that condition (7.20) is satisfied. Then*

(i) *the function h is injective on Δ and $h(\Delta) = \Omega$ is a convex domain,*

(ii) *for any $n \geq 2$ and any $z \in \Delta$ the following sharp estimate is valid:*

$$\left| \frac{h^{(n)}(z)}{h'(z)} - \frac{n\bar{z}}{1 - |z|^2} \frac{h^{(n-1)}(z)}{h'(z)} \right| \leq \frac{n!}{(1 - |z|)^{n-1}(1 + |z|)}. \tag{7.21}$$

Proof. The condition (7.20) immediately implies $h(t) \neq h(z)$ for $z \in \Delta$, $t \in \Delta$, $t \neq z$, and therefore the injectivity of the function h on Δ. Now, we fix $t \in \Delta$ and we consider the function

$$\varphi(z) = 2 \frac{h(z) - h(t)}{h'(t)} \frac{1 - \bar{t}z}{(z - t)(1 - |t|^2)} - 1, \quad z \in \Delta.$$

It is evident that $\varphi(t) = 1$ and that $\operatorname{Re}\varphi(z) > 0$ for any $z \in \Delta$. In a neighbourhood of the point t we have the Taylor expansion

$$\varphi(z) = 1 + 2 \sum_{n=2}^{\infty} \left(\frac{h^{(n)}(t)}{h'(t)n!} - \frac{\bar{t}}{(1 - |t|^2)(n - 1)!} \frac{h^{(n-1)}(t)}{h'(t)} \right) (z - t)^{n-1}.$$

Since $1/\lambda_\Lambda(\varphi(t)) = 2 \operatorname{Re}\varphi(t) = 2$, using the Schwarz-Pick lemma one easily gets

$$|\varphi'(t)| = \left| \frac{h''(t)}{h'(t)} - \frac{2\bar{t}}{1 - |t|^2} \right| \leq \frac{2}{1 - |t|^2}, \quad t \in \Delta,$$

which is equivalent to the inequality

$$\left| w - \frac{1 + |t|^2}{1 - |t|^2} \right| \leq \frac{2|t|}{1 - |t|^2}, \quad t \in \Delta, \tag{7.22}$$

where

$$w = 1 + t \frac{h''(t)}{h'(t)}.$$

The condition (7.22) implies $\operatorname{Re} w > 0$. Therefore (see for instance [128] and [146]), h is injective on Δ and $\Omega = h(\Delta)$ is a convex domain.

To get (ii) for $n \geq 3$ we apply Ruscheweyh's theorem 4.6 to get sharp estimates for the derivatives

$$\frac{\varphi^{(n-1)}(t)}{(n - 1)!} = 2 \left(\frac{h^{(n)}(t)}{nh'(t)} - \frac{n\bar{t}}{1 - |t|^2} \frac{h^{(n-1)}(t)}{h'(t)} \right), \quad t \in \Delta,$$

indicated in (ii). Equality in (7.21) at the point $z = z_0 \in \Delta$ occurs if

$$h(z) = \frac{z}{1 - \overline{z_0}z/z_0} = \frac{z z_0}{z_0 - \overline{z_0}z}.$$

This completes the proof of Theorem 7.18. $\qquad \square$

Remark 7.19. According to the above, the condition (7.20) is a new necessary and sufficient condition for $h(\Delta)$ to be convex that does not use the second derivative of h. It may be worthwhile to mention the conditions of this type that have been proved before. To our knowledge the first one was

$$\text{Re}\left(\frac{zh'(z)}{h(z) - h(t)}\right) > 0, \quad |t| < |z| < 1,$$

proved by Brickman in [46]. Sheil-Small ([151]) and Suffridge ([156]) proved the characterization

$$\text{Re}\left(\frac{zh'(z)}{h(z) - h(t)} - \frac{t}{z - t}\right) > \frac{1}{2}, \quad z \in \Delta, \ t \in \Delta,$$

of convex funtions h. The third condition we want to cite seems to be the most famous. It has been proved by Ruscheweyh in [141] and it was used by Ruscheweyh and Sheil-Small in [144] to prove the Pólya-Schoenberg conjecture. This one is as follows.

$$\text{Re}\left(\frac{z}{z - \zeta} \frac{\zeta - t}{z - t} \frac{h(z) - h(t)}{h(\zeta) - h(t)} - \frac{\zeta}{z - \zeta}\right) > \frac{1}{2}, \quad z \in \Delta, \ t \in \Delta, \ \zeta \in \Delta.$$

One curious difference between these characterizations of convexity of $h(\Delta)$ and (7.20) seems to be that (7.20) contains nonanalytic terms.

Chapter 8

Some open problems

8.1 The Krzyż conjecture

As we have seen in Chapter 6 it is a very difficult task to find sharp punishing factors or substitutes for them in cases where multiply connected domains are involved. From this point of view, it seems natural that the difficulties become nearly insuperable, if one allows the points $z_0 \in \Omega$ or $f(z_0) \in \Pi$, or both to vary, and asks for the maximum. Nevertheless, there exists one problem of this type that has attracted researchers for many years because of the conjectured simple solution. This is the so-called Krzyż conjecture.

It is concerned with functions $f \in A(\Delta, \Delta')$, where

$$\Delta' = \{z \mid 0 < |z| < 1\}$$

and the conjecture is that

$$\max\left\{ \frac{|f^{(n)}(0)|}{n!} \mid f \in A(\Delta, \Delta') \right\} = \frac{2}{e}$$

for any $n \in \mathbb{N}$ (see [96]). It is obvious that the conjectured bound is attained for the functions

$$f(z) = \exp\left(\frac{1 + z^n}{1 - z^n} \right).$$

The story of this conjecture began with a problem posed in [101], where the case $n = 1$ of the above is considered. It is remarkable that in the solutions found in 1934 the universal covering function of Δ' plays the decisive role. After the formulation of the conjecture by Krzyż, there was a lot of effort to solve this attractive problem. The cases $n = 2$ and $n = 3$ were solved in [84], for $n = 4$ see [161], [47], and [158]. The case $n = 5$ was solved recently in [145].

We cannot deny that our considerations for the case of multiply connected domains began with the hope to get a simple dependence of the upper bound of

$f(0)$ in this difficult problem. We were able with the above theorems to get the simple cases $n = 1$ and $n = 2$, but not more.

We may be allowed to conclude this section with a little observation which may prevent others from attacking this problem. In the above cited article, Szapiel poses the following problem: *Do the estimates*

$$\left| \sum_{j=1}^{n} c_j A_j(F) \right| \leq n, \tag{8.1}$$

hold for all schlicht functions

$$F(z) = z + \sum_{n=2}^{\infty} A_n(F) z^n,$$

whenever

$$\left| \sum_{j=1}^{n} c_j \zeta^{j-1} \right| \leq 1 \quad \text{for all } |\zeta| \leq 1? \tag{8.2}$$

We find that the answer is positive and that more is true.

Theorem 8.1. *Let c_j satisfy the condition* (8.2). *Then the inequality* (8.1) *holds for any function F subordinate to a schlicht function $F_0 \in S$.*

Proof. If we define

$$P(z) = \sum_{j=1}^{n} c_j \zeta^j,$$

we see that the Schwarz lemma indicates that the condition (8.2) is equivalent to $P(\overline{\Delta}) \subset \overline{\Delta}$. Therefore the Sheil-Small theorem (see Chapter 2 and [152]) imply that the inequality

$$|(P * F)(e^{i\theta})| \leq n$$

holds for all $\theta \in [0, 2\pi]$ and not only for all $F \in S$, but even for the case that F is subordinated to a schlicht function. This proves the theorem, a far-reaching generalization of (8.1). □

8.2   The angle conjecture

As we have seen in Chapter 5, for $f \in A(\Omega, H_\alpha)$, $\alpha \in [1, 2]$, the inequalities

$$\frac{|f^{(n)}(z_0)|}{n!} \leq \frac{2^{n-1}}{\alpha} \binom{n + \alpha - 1}{n} \frac{(\lambda_\Omega(z_0))^n}{\lambda_\Pi(f(z_0))}, \quad n \geq 2, \ z_0 \in \Omega,$$

are valid as well for $\Omega = \Delta$ as for $\Omega = H_1$. It seems reasonable that these inequalities are true for any convex domain Ω. Concerning this conjecture, we will show now that it is true, if $f(z_0)$ lies on the line bisecting the angle H_α.

Theorem 8.2. *Let Ω be a convex proper subdomain of $\mathbb{C}$ and H_α an angular domain with opening angle $\alpha\pi$, $\alpha \in [1,2]$. Let further $f \in A(\Omega, H_\alpha)$, $z_0 \in \Omega$, and $f(z_0)$ lying on the bisector of H_α. Then for $n \geq 2$ the inequalities*

$$\frac{|f^{(n)}(z_0)|}{n!} \leq \frac{2^{n-1}}{\alpha} \binom{n+\alpha-1}{n} \frac{(\lambda_\Omega(z_0))^n}{\lambda_\Pi(f(z_0))}$$

are valid.

Proof. Let v be the vertex of the angular domain H_α and

$$f(z_0) - v = |f(z_0) - v|e^{i\varphi_0}.$$

Then the Riemann mapping function $\Phi_{H_\alpha, f(z_0)}$ is given by

$$\Phi_{H_\alpha, f(z_0)}(\zeta) = v + (f(z_0) - v)\left(\frac{1 + ze^{-i\varphi_0}}{1 - ze^{-i\varphi_0}}\right)^\alpha.$$

Hence,

$$\lambda_{H_\alpha}(f(z_0)) = \frac{1}{2\alpha|f(z_0) - v|}.$$

Choosing an appropriate function $\omega : \Delta \to \overline{\Delta}$ such that

$$\tilde{\Phi}(\zeta) = \Phi_{H_\alpha, f(z_0)}(\zeta\omega(\zeta))$$

has the property $\tilde{\Phi}(\Delta) = f(\Omega)$, we get

$$f = \tilde{\Phi} \circ \Psi_{\Omega, z_0}.$$

If we now proceed as in the above proofs of the generalizations of the two Chua conjectures, we see that for the proof of the present assertion we have to show that

$$\left|\sum_{k=1}^{n} c_k(\alpha)A_{n,k}\right| \leq \frac{2^{n-1}}{\alpha}\binom{n+\alpha-1}{n}, \tag{8.3}$$

where the function represented by the series

$$\sum_{k=1}^{\infty} c_k(\alpha)\zeta^k$$

is subordinated to the function

$$\frac{1}{2\alpha}\left(\left(\frac{1+\zeta}{1-\zeta}\right)^\alpha - 1\right) = \sum_{k=1}^{\infty} h_k(\alpha)\zeta^k \tag{8.4}$$

and the $A_{n,k}$ represent the coefficients of the powers of inverses of convex functions as above.

Analogous to the proofs mentioned above we want to show now that for functions

$$\omega(z) = \sum_{\tau=0}^{\infty} d_\tau \zeta^k,$$

mapping the unit disc into the closed unit disc, the inequalities

$$\left| \sum_{\tau=0}^{p-1} (p-\tau)c_{p-\tau}(\alpha)d_\tau \right| \leq p\, h_p(\alpha), \quad p \in \mathbb{N}, \tag{8.5}$$

are valid. To this end, we consider, as in the proof of the 2^{n-1} conjecture, the region of variability of the linear functional

$$L_\omega(g) = \sum_{\tau=0}^{p-1} (p-\tau)c_{p-\tau}(\alpha)d_\tau,$$

where ω is fixed and g is varying in the family of functions subordinated under (8.4). According to [44], [45], and [146] these functions g have a representation

$$g(\zeta) = \frac{1}{2\alpha} \int_0^{2\pi} \left(\left(\frac{1+e^{it}\zeta}{1-e^{it}\zeta} \right)^\alpha - 1 \right) d\mu(t),$$

with a probability measure $\mu : [0, 2\pi] \to \mathbb{R}$. Hence, it is sufficient to prove the inequality (8.5) for $c_\tau(\alpha) = e^{i\tau} h_\tau(\alpha)$. Further, the rotational invariance of the family of unimodular bounded functions indicates that we only need to show that

$$\left| \sum_{\tau=0}^{p-1} (p-\tau)h_{p-\tau}(\alpha)d_\tau \right| \leq p\, h_p(\alpha), \quad p \in \mathbb{N}. \tag{8.6}$$

We want to thank at this point St. Ruscheweyh for pointing out to us the idea of this proof. His idea was to consider the polynomial

$$Q(\zeta) = \sum_{\tau=0}^{p-1} \frac{(p-\tau)h_{p-\tau}(\alpha)}{p\, h_p(\alpha)} \zeta^k$$

and to show that the function $Q * \omega$ is a unimodular bounded function.

According to [152] (see especially the theorems (2.6) and (3.2) of this article), this can be reduced to the proof of

$$\mathrm{Re}(Q(\zeta)) > \frac{1}{2}, \quad \zeta \in \Delta.$$

By the application of a theorem due to Rogosinski in [138], to achieve this aim it is sufficient to prove that the coefficients of Q form a monotonic decreasing convex sequence. This means that

$$(k+1)h_{k+1}(\alpha) - k\, h_k(\alpha) \geq 0, \quad k = 1, \ldots, p-1,$$

$$(k+2)h_{k+2}(\alpha) - 2(k+1)h_{k+1}(\alpha) + k\,h_k(\alpha) \geq 0, \quad k = 1, \ldots, p-2,$$

and

$$2h_2(\alpha) - 2h_1(\alpha) \geq 0.$$

To prove these inequalities we have to insure that the coefficients of

$$\frac{1}{2\alpha}\frac{d}{d\zeta}\left(\frac{1+\zeta}{1-\zeta}\right)^\alpha = \frac{(1+\zeta)^{\alpha-1}}{(1-\zeta)^{\alpha+1}}$$

form a monotonic increasing convex sequence. This is easily seen by recognizing that

$$\frac{(1+\zeta)^{\alpha-1}}{(1-\zeta)^{\alpha+1}}(1-\zeta)^2 = e^{(\alpha-1)\log\left(\frac{1+\zeta}{1-\zeta}\right)}$$

has nonnegative coefficients.

A comparison with the proofs of the generalizations of the two conjectures of Chua above reveals that we can get the inequality (8.3), setting again

$$(\omega(\zeta))^\sigma = \sum_{j=0}^{\infty} d_{j,\sigma}\zeta^j, \quad \zeta \in \Delta$$

and using (8.6) in the chain

$$\left|\sum_{k=1}^{n} c_k(\alpha)A_{n,k}\right| \leq |c_n(\alpha)| + \sum_{\sigma=1}^{n-1}\frac{1}{n}\binom{n}{\sigma}\left|\sum_{j=\sigma}^{n-1}(n-j)c_{n-j}(\alpha)d_{j-\sigma,\sigma}\right|$$

$$\leq h_n(\alpha) + \sum_{\sigma=1}^{n-1}\frac{1}{n}\binom{n}{\sigma}(n-\sigma)h_{n-\sigma}(\alpha)$$

$$= \sum_{k=1}^{n}\binom{n-1}{n-k}h_k(\alpha) = \frac{2^{n-1}}{\alpha}\binom{n+\alpha-1}{n}.$$

The last identity has been shown in Chapter 5. □

8.3 The generalized Goodman conjecture

In many of the results of the present book it could be proved that the bounds for the moduli of the Taylor coefficients of the functions in a certain family of functions holomorphic in Δ are the right bounds for the related coefficients of the functions subordinated to the above ones. Therefore, we dare to formulate a challenging conjecture that generalizes as well the Rogosinski conjecture, which naturally is a theorem nowadays, as Jenkins' theorem 2.7 on the Goodman conjecture.

The generalized Goodman conjecture. Let f be injective and meromorphic in Δ. Further let f be normalized by $f(0) = f'(0) - 1 = 0$ and have the pole at the point $p \in (0,1)$. If $\omega : \Delta :\to \overline{\Delta}$, then we conjecture that for the expansion

$$g(z) = f(z\omega(z)) = \sum_{n=1}^{\infty} a_n(g)z^n,$$

valid in some neighbourhood of the origin, the inequalities

$$|a_n(g)| \leq \frac{1}{p^{n-1}} \sum_{k=0}^{n-1} p^{2k}, \quad n \geq 2,$$

are valid.

This relation between f and g will be denoted by $g \prec f$ as in the holomorphic case. To support this conjecture we prove that it is valid in some special cases. The first one deals with little values of p.

Theorem 8.3. *The generalized Goodman conjecture is valid for $n = 2$, $p \in (0,1)$, for $n = 3$, $p \in (0, 0.7)$, and for $n \geq 4$, $p \in (0, 1/(2n-2))$.*

Proof. For $f(z) = z + \sum_{n=2}^{\infty} a_n z^n$, $|z| < p$ meromorphic and univalent in D with pole $p \in (0,1)$ and $w(z) = \sum_{n=0}^{\infty} c_n z^n$, $\omega : D \to \overline{D}$, we have to find the least upper bound for

$$|c_1 + a_2(c_0)^2|$$

in the case $n = 2$ and for

$$|c_2 + 2c_0 c_1 a_2 + (c_0)^3 a_3|$$

in the case $n = 3$. In the first the triangle inequality, Jenkins' inequality (see [86]), and the Schur algorithm (see[148]) deliver as an upper bound

$$1 - |c_0|^2 + |c_0|^2 \frac{1 + p^2}{p} \leq \frac{1 + p^2}{p}.$$

In the second case, we use

$$\left| \frac{c_2}{1 - |c_0|^2} + \frac{\overline{c_0}(c_1)^2}{(1 - |c_0|^2)^2} \right| \leq 1 - \frac{|c_1|^2}{(1 - |c_0|^2)^2}$$

(see[148]). Using the triangle inequality and the abbreviations

$$x = |c_0|, \quad y = \frac{|c_1|}{(1 - |c_0|^2)}$$

we get

$$|c_2| \leq (1 - x^2)(xy^2 + 1 - x^2).$$

Therefore, we have to maximize

$$(1-x^2)(xy^2 + 1 - x^2) + 2xy(1-x^2)\frac{1+p^2}{p} + x^3\frac{1+p^2+p^4}{p^2} = F(x,y;p)$$

for $(x,y) \in [0,1] \times [0,1]$.

Since

$$\frac{\partial F}{\partial y} = 2(1-x^2)\left(y(x-1) + x\frac{1+p^2}{p}\right) \geq (\leq) 0$$

for

$$y \leq (\geq)\frac{x}{1-x}\frac{1+p^2}{p},$$

and

$$\frac{x}{1-x}\frac{1+p^2}{p} \leq 1$$

for

$$x \leq \frac{p}{1+p+p^2},$$

in these cases we have to check the local maximum of F in

$$y = \frac{x}{1-x}\frac{1+p^2}{p}.$$

Inserting this delivers a monotonic increasing function of x. So we have to check that this function at

$$x = \frac{p}{1+p+p^2}$$

is less than $(1+p^2+p^4)/p^2$. This is the case for $p \in (0,1)$.

For

$$x \geq \frac{p}{1+p+p^2}$$

we have to consider

$$F(x,1,p) = x\frac{2+p+2p^2}{p} + x^3\frac{1-2p-2p^3+p^4}{p}.$$

For those p for which the derivative of this function with respect to x is positive for $x \leq 1$, the proof is complete.

This is the case for

$$p \leq \sqrt{\frac{2+\sqrt{19}}{3}} = 0.7088023\dots.$$

For $n \geq 4$ we consider

$$(z\omega(z))^k = \sum_{j=k}^{\infty} c_{j,k} z^j$$

and we see that $c_{k,k} = (c_0)^k$ and therefore

$$|c_{j,k}| \leq 1 - |c_0|^{2k}, \quad j \geq k.$$

Hence,

$$|a_n(g)| = \left| \sum_{k=1}^{n} c_{n,k} a_k \right| \leq \sum_{k=1}^{n-1} \left((1 - |c_0|^{2k}) \sum_{j=0}^{k-1} \frac{p^{2j}}{p^{k-1}} \right) + |c_0|^n \sum_{j=0}^{n-1} \frac{p^{2j}}{p^{n-1}}.$$

We want to prove that this is less than or equal to $\sum_{j=0}^{n-1} (p^{2j}/p^{n-1})$. This inequality is equivalent to

$$\sum_{k=1}^{n-1} \left(\sum_{j=0}^{2k-1} |c_0|^j \sum_{j=0}^{k-1} \frac{p^{2j}}{p^{k-1}} \right) \leq \sum_{j=}^{n-1} |c_0|^j \sum_{j=0}^{n-1} \frac{p^{2j}}{p^{n-1}}.$$

We use the rough estimate

$$\sum_{j=0}^{2k-1} |c_0|^j \leq 2 \sum_{j=0}^{n-1} |c_0|^j, \quad 1 \leq k \leq n - 1,$$

and we see by some elementary calculations that

$$2 \sum_{k=1}^{n-1} \sum_{j=0}^{k-1} \frac{p^{2j}}{p^{k-1}} \leq \sum_{j=0}^{n-1} \frac{p^{2j}}{p^{n-1}}$$

for $p \in (0, 1/(2n - 2))$. $\square$

The second theorem is concerned with the condition that $f(\Delta)$ is concave, i.e., $\overline{\mathbb{C}} \setminus f(\Delta)$ is a convex compact set. The family of functions with these properties will be denoted by $Co(p)$ as in Chapter 5, Section 1, and the family of functions $\omega : \Delta \to \overline{\Delta}$ by B.

Theorem 8.4 (see [26]). *Let $p \in (0, 1)$. If $f \in Co(p)$ and $g \prec f$, then*

$$|a_n(g)| \leq \frac{1}{p^{n-1}} \sum_{k=0}^{n-1} p^{2k}, \quad n \in \mathbb{N}. \tag{8.7}$$

For the proof of Theorem 8.4 we use a representation formula for the functions in the class $Co(p)$ that was derived in [167] and [23] as a simple consequence of Theorem 4 in [117].

Theorem 8.5. *Let $p \in (0, 1)$. For any $f \in Co(p)$ there exists a function $w_1 \in B$ such that*

$$f(z) = \frac{z}{1 - \frac{z}{p}} \frac{1 - \frac{p}{1+p^2}(1 + w_1(z))z}{1 - zp}, \quad z \in \Delta. \tag{8.8}$$

The first step in the proof of Theorem 8.4 concerns the first factor in the representation (8.8) that we will denote by f_1,

$$f_1(z) = \frac{z}{1 - \frac{z}{p}}, \quad z \in \Delta.$$

We prove that, for this member of $Co(p)$, one may replace the sum in Theorem 8.4 by 1.

Theorem 8.6. *Let* $p \in (0,1)$. *If* $g \prec f_1$, *then* $|a_n(g)| \leq \frac{1}{p^{n-1}}$, $n \in \mathbb{N}$.

Proof. Let $w \in B$ such that $g(z) = f_1(zw(z))$, $z \in \Delta$. This implies

$$g(z) = w(z) \left(z + \frac{zg(z)}{p} \right), \quad z \in \Delta. \tag{8.9}$$

Now, to prove Theorem 8.6, we use a method due to Clunie (see [57]) and a generalization of a theorem of Robertson (see [137]) proved in [30], Lemma 2.1. This is the meromorphic version of the Rogosinski lemma (see our Theorem 2.4).

The application of this lemma to the identity (8.9) yields that for any $n \in \mathbb{N}$, the inequality

$$\sum_{k=1}^{n} |a_k(g)|^2 \leq 1 + \sum_{k=2}^{n} \frac{|a_{k-1}(g)|^2}{p^2}$$

is valid. The assertion of Theorem 8.6 follows by mathematical induction using $1/p > 1$. $\square$

Next, we consider the second factor in the representation (8.8) that we will denote by f_2 and we prove the following inequalities.

Theorem 8.7. *Let* $p \in (0,1)$ *and for* $w \in B$,

$$g_2(z) = f_2(zw(z)) = 1 + \sum_{k=1}^{\infty} B_k z^k, \quad z \in D.$$

Then for any $n \in \mathbb{N}$ *the inequality*

$$\sum_{k=1}^{n} |B_k|^2 \leq \sum_{k=1}^{n} p^{2k} \tag{8.10}$$

is valid.

Proof. Obviously,

$$f_2(z) = 1 + \frac{\frac{p^2 - w_1(z)}{1 + p^2} \, p \, z}{1 - zp}, \quad z \in \Delta.$$

Since

$$\left| \frac{p^2 - w_1(z)}{1 + p^2} \right| \leq 1, \quad z \in \Delta,$$

there exists a function $w_2 \in B$ such that

$$f_2(z) = 1 + \frac{w_2(z)\,p\,z}{1 - zp}, \quad z \in \Delta.$$

Let $w_3(z) = w_2(zw(z))$, $z \in \Delta$. Then $w_3 \in B$. Hence, the function $g_2 - 1$,

$$g_2(z) - 1 = \frac{w_3(z)\,p\,zw(z)}{1 - zw(z)p}, \quad z \in \Delta,$$

is quasi-subordinate to the function h,

$$h(z) = \frac{pz}{1 - zp}, \quad z \in \Delta.$$

The notation of quasi-subordination was defined by Robertson (see [137]). He proved that the above relation implies in our case the inequality (7.9) (compare also [128], Theorem 2.2). This completes the proof of Theorem 8.5.   □

After these preparations, it is easy to prove Theorem 8.4 as follows.

Proof of Theorem 8.4. If $g \prec f$ and $f \in Co(p)$, we conclude from Theorem 8.5 that there exists a function $w \in B$ such that

$$g(z) = f_1(zw(z))\,f_2(zw(z)). \tag{8.11}$$

Let

$$f_1(zw(z)) = \sum_{k=1}^{\infty} A_k z^k,$$

where this expansion is valid in some neighbourhood of the origin. From (8.11) we get for $n \in \mathbb{N}$,

$$a_n(g) = A_n + \sum_{k=1}^{n-1} A_k B_{n-k}.$$

The Cauchy-Schwarz inequality shows that, for any $n \in \mathbb{N}$, the inequality

$$|a_n(g)| \leq \left(\sum_{k=1}^{n} |A_k|^2 \right)^{1/2} \left(1 + \sum_{k=1}^{n-1} |B_k|^2 \right)^{1/2}$$

is valid. If we use Theorems 8.6 and 8.7 to estimate the right-hand side of this inequality, we get

$$|a_n(g)| \leq \left(\sum_{k=0}^{n-1} p^{-2k} \right)^{1/2} \left(\sum_{k=0}^{n-1} p^{2k} \right)^{1/2} = \frac{1}{p^{n-1}} \sum_{k=0}^{n-1} p^{2k}.$$

This completes the proof of Theorem 8.4.   □

It is obvious that the estimates in Theorems 8.4 and 8.6 are sharp. In Theorem 8.6, equality is attained if

$$g(z) = \frac{cz}{1 - \frac{cz}{p}}, \quad |c| = 1.$$

In Theorem 8.4 equality occurs for

$$g(z) = \frac{cz}{\left(1 - \frac{cz}{p}\right)(1 - zp)}, \quad |c| = 1.$$

These are conformal maps of Δ onto

$$\overline{\mathbb{C}} \setminus \left[\frac{-p}{(1-p)^2}, \frac{-p}{(1+p)^2} \right]$$

(compare [15] and [166]).

The third theorem proves the generalized Goodman conjecture for univalent functions with real coefficients. They belong to the class of functionss typically-real and meromorphic, introduced by Goodman in [72]. The defining relation for this class is, as in the holomorphic case,

$$\Im(f(z)\,\Im(z) \geq 0, \quad z \in \Delta.$$

It is easily seen that this implies that in our case the residuum of f at the point p is negative. Goodman proved a representation theorem that implies the following application to univalent meromorphic and typically-real functions normalized as usual.

Theorem 8.8 (see [72] Theorem 9). *Let f be a function univalent meromorphic and typically-real in Δ having the residuum $-m < 0$ at the point p. There exists a function t holomorphic and typically-real in Δ and a nonnegative constant M such that*

$$M + m\left(\frac{1}{p} - 1\right) = 1$$

and

$$f(z) = M\,t(z) + m\left(\frac{1}{p} - 1\right)\frac{z}{(1 - zp)(1 - z/p)}, \quad z \in \Delta.$$

Using this representation it is easy to prove (8.7) for univalent meromorphic and typically-real functions.

Theorem 8.9. *Let f be univalent meromorphic and typically-real in Δ. If f has its pole at the point $p \in (0,1)$ and $g \prec f$, then (8.7) holds.*

Proof. According to a theorem proved by Robertson in [135], for any function typically-real and holomorphic in Δ there exists a probability measure μ on $[0, \pi]$ such that

$$t(z) = \int_0^\pi \frac{z \, d\mu(\theta)}{1 - 2z \cos \theta + z^2}, \quad z \in \Delta.$$

Since the kernel functions belong to the class S, the truth of the Rogosinski conjecture for univalent functions and a convex hull argument imply that for $g_1 \prec t$ the inequalities $|a_n(g_1)| \le n$, $n \ge 2$ hold.

From Theorem 8.4 above we know that for

$$g_2(z) \prec \frac{z}{(1 - zp)(1 - z/p)},$$

the generalized Goodman conjecture holds. Hence,

$$|a_n(g_2)| \le \frac{1}{p^{n-1}} \sum_{k=0}^{n-1} p^{2k}.$$

From the identity

$$\frac{1}{p^{n-1}} \sum_{k=0}^{n-1} p^{2k} = \begin{cases} \sum_{k=1}^u \left(\frac{1}{p^{2k-1}} + p^{2k-1} \right), & \text{if } n = 2u, \\ 1 + \sum_{k=1}^u \left(\frac{1}{p^{2k}} + p^{2k} \right), & \text{if } n = 2u+1 \end{cases}$$

we conclude that

$$\frac{1}{p^{n-1}} \sum_{k=0}^{n-1} p^{2k} \ge n.$$

If we write a function $g \prec f$ in the form

$$g(z) = Mg_1(z) + m \left(\frac{1}{p} - 1 \right) g_2(z), \quad z \in \Delta,$$

we see that

$$|a_n(g)| \le M|a_n(g_1)| + m \left(\frac{1}{p} - 1 \right) |a_n(g_2)| \le \frac{1}{p^{n-1}} \sum_{k=0}^{n-1} p^{2k}.$$

This proves Theorem 8.9. $\square$

Using the bound found by Jenkins and well-known subordination techniques, it is not difficult to find upper bounds for $|a_n(g)|$. We give two examples

$$|a_n(g)| \le \sum_{k=1}^n \left(\frac{1}{p^{k-1}} \sum_{j=0}^{k-1} p^{2j} \right)$$

and

$$|a_n(g)| \leq \left(\sum_{k=1}^{n} \left(\frac{1}{p^{k-1}} \sum_{j=0}^{k-1} p^{2j} \right)^2 \right)^{1/2}.$$

It seems that these estimates reflect approximately the right asymptotics of the least upper bounds for $p \to 0$, but it is evident that they are bad for $p \to 1$.

An upper bound that fits better for the neighbourhood of $p = 1$ can be found if one applies the validity of the Rogosinski conjecture to the function $f_p(z) = f(pz)/p$ that belongs to the class S. By this procedure we get

$$|a_n(g)| \leq \frac{n}{p^{n-1}}.$$

We add a last inequality for the coefficients $a_n(g)$ that follows immediately from the generalized Goluzin-Rogosinski theorem 2.5 using Jenkins' theorem 2.7. Taking

$$\lambda_k = \left(\frac{p^{n-1}}{\sum_{j=0}^{k-1} p^{2j}} \right)^2, \quad k \in \mathbb{N},$$

we get

$$\sum_{k=1}^{n} |a_k(g)|^2 \lambda_k \leq \sum_{k=1}^{n} |a_k|^2 \lambda_k \leq n.$$

Obviously, this inequality is sharp for the function for which Jenkins' theorem is sharp.

8.4 Bloch and several variable problems

We want to avoid the impression that the above considerations are the only possibility to generalize the Schwarz-Pick lemma to higher derivatives. Therefore we mention some questions that arose when we studied the work of colleagues on generalized Schwarz-Pick type estimates (see [38], [112], and [113]).

For example, it has been proved in [112] that for $f \in A(\Delta, \Delta)$, $\alpha, \beta > 0$, the implications

$$\sup_{z \in \Delta} \frac{|f'(z)|(1 - |z|^2)^\beta}{(1 - |f(z)|^2)^\alpha} < \infty$$

$$\Longrightarrow \sup_{z \in \Delta} \frac{|f^{(n)}(z)|(1 - |z|^2)^{\beta+n-1}}{(1 - |f(z)|^2)^\alpha} < \infty, \quad n \geq 2,$$

are valid.

On one hand it seems natural to ask whether one can compute an explicit relation between these two suprema. On the other hand this result together with our theorems on punishing factors suggests the question for which pairs (Ω, Π) of domains and $\alpha, \beta > 0$ implications of the form

$$\sup_{z \in \Omega} \frac{|f'(z)| R(\Omega, z)^\beta}{R(\Pi, f(z))^\alpha} < \infty$$

$$\Longrightarrow \sup_{z \in \Omega} \frac{|f^{(n)}(z)| R(\Omega, z)^{\beta + n - 1}}{R(\Pi, f(z))^\alpha} < \infty, \quad n \geq 2,$$

are valid.

The results of MacCluer, Stroethoff, and Zhao in [113] may give rise to analogous problems even for functions of several variables. Since it would be lengthy to cite their definitions and theorems, we prefer to conclude with open questions that are simpler to formulate and that originated in the following result of Bénéteau, Dahlner, and Khavinson (see [38]). They proved that for an analytic function

$$f : \Delta^n \to \Delta, \quad (z_1, \ldots, z_n) \mapsto f(z_1, \ldots, z_n)$$

and any multiindex

$$\alpha = (\alpha_1, \ldots, \alpha_n) \in (\mathbb{N} \cup \{0\})^n$$

the inequality

$$\sup_{(z_1, \ldots, z_n) \in \Delta^n} \frac{|D^\alpha f(z_1, \ldots, z_n)| \prod_{k=1}^n \left(1 - |z_k|^2\right)^{\alpha_k}}{1 - |f(z)|^2} < \infty$$

holds, where $D^\alpha f(z_1, \ldots, z_n)$ is the usual abbreviation for the derivative of order α.

Again, one may ask for estimates for this supremum or consider generalizations of this result to other domains instead of Δ.

For new results on similar generalizations of Schwarz-Pick estimates compare [5], [6], [68] and [92].

8.5   On sums of inverse coefficients

As we have seen frequently above, a decisive part in the computation of punishing factors is played by inverse coefficients $A_{n,k}$ of the kth powers of functions injective in the unit disc. Löwner's Theorem 2.7 and its generalizations provided us with satisfactory solutions for these problems. If subclasses of S are considered, the situation is far less nice. The most famous of such problems is the question for bounds in the convex case.

For the function

$$f(z) = \frac{z}{1 - cz}, \quad |c| = 1, \tag{8.12}$$

the identities

$$|A_{n,k}| = \binom{n-1}{k-1}$$

hold for $1 \leq k \leq n \in \mathbb{N}$.

R. L. Libera and E. J. Zlotkiewicz proved in [104] that

$$|A_{n,1}| \leq 1, \quad n = 2, 3, 4, 5, 6, 7$$

for the inverse coefficients of convex functions. This result implies the validity of the estimate

$$|A_{n,k}| \leq \binom{n-1}{k-1} \tag{8.13}$$

for n between 2 and 7.

On the other hand W. E. Kirwan and G. Schober got by explicit computations that

$$M_{10} = \max\{|A_{10,1}|\} > 1.2,$$

where the maximum is taken over all convex functions (see [90]). A detailed discussion of

$$M_n = \max\{|A_{n,1}|\}$$

may be found in [50] and [58]. J.T.P. Campschroer showed in [50] that

$$M_n = O\left(2^n n^{-3}\right).$$

It may be interesting that for k big compared with n the inequalities (8.13) hold true.

Theorem 8.10. *For the inverse coefficients $A_{n,k}$ of convex functions, $n \geq 4$, and $n/2 - 1 \leq k \leq n$, the inequalities (8.13) are valid.*

Proof. As above, we use the Schur-Jabotinsky theorem and, in addition, the fact that a convex function f is starlike of order $1/2$. This implies that the function h defined by

$$h(z) = z \left(\frac{f(z)}{z}\right)^2$$

is a starlike function. Hence, we may consider the Taylor coefficients of

$$\left(\frac{z}{h(z)}\right)^{n/2} = \sum_{n=0}^{\infty} b_{l,n/2}(h) z^n$$

where h is starlike. But from [140] we know that the modulus $|b_{l,n/2}(h)|$ of the Taylor coefficients of these functions is bounded from above by $|b_{l,n/2}(k)|$, where k is the Koebe function or one of its rotations and $l \leq n/2 + 1$. In most cases, the Koebe functions are the only extremal functions. This proves our theorem. $\square$

But in our computations of punishing factors, weighted sums of the $A_{n,k}$ are the most important elements. If one looks at punishing factors for convex pairs (see section 4.5), one may recognize that the proof of Theorem 4.12 would be very short, if the following conjecture (compare [22]) could be proved:

Conjecture. *For the inverse coefficients of convex functions and for any $n \geq 2$,
the inequality*

$$\sum_{k=1}^{n} |A_{n,k}| \leq 2^{n-1}$$

is valid.

The validity of this conjecture for $2 \leq n \leq 7$ follows immediately from the
above.

A sharp estimate for the sum of squares can be derived from section 4.5 and
an application of Theorem 2.5 on subordinate functions. Actually, one easily gets

$$\sum_{k=1}^{n} |A_{n,k}|^2 = \sum_{\mu=0}^{n-1} \left| \frac{n-\mu}{n} \, a_{\mu,n} \right|^2$$

$$\leq \sum_{\mu=0}^{n-1} \left(\frac{n-\mu}{n} \right)^2 \binom{n}{\mu}^2 = \binom{2(n-1)}{n-1}.$$

Here, equality is attained for the functions (8.12).

If one uses this inequality to estimate the sum of moduli of the inverse coefficients by the Cauchy inequality and the Stirling formula, one gets immediately

$$\sum_{k=1}^{n} |A_{n,k}| \leq \sqrt{n} \left(\frac{2(n-1)}{n-1} \right)^{1/2} \leq 2^{n-1} n^{1/4}.$$

If one wants to compute punishing factors for functions holomorphic on β-
accessible domains, it would be necessary to know something of the inverse coefficients of functions close-to-convex of order β or of weighted sums of the moduli of
these inverse coefficients (compare section 7.1 and section 5.6 on lower bounds).
The problem of such inverse coefficients is addressed in [90], but those results do
concern only early coefficients. It seems natural to us to conjecture inequalities for
the sums mentioned above to get new conjectures on punishing factors.

Bibliography

[1] S. Agard, *Distortion theorems for quasiconformal mappings*. Ann. Acad. Sci. Fenn. Ser. AI **413** (1968), 12p.

[2] D. Aharonov and S. Friedland, *On an inequality connected with the coefficient conjecture for functions of bounded boundary rotation*. Ann. Acad. Sci. Fenn. Ser. AI **524** (1972), 14p.

[3] L. V. Ahlfors, *Conformal invariants, Topics in Geometric Function Theory*. McGraw-Hill, New York, 1973.

[4] A. Ancona, *On strong barriers and an inequality of Hardy for domains in* $\mathbb{R}^n$. J. London Math. Soc.(2) **34** (1986), no.2, 274–290.

[5] J. M. Anderson, M. A. Dritschel, and J. Rovnyak, *Schwarz-Pick inequalities for the Schur-Agler class on the polydisk and unit ball*. Comp. Methods and Function Theory **8** (2008), 339–361.

[6] J. M. Anderson and J. Rovnyak, *On generalized Schwarz-Pick estimates*. Mathematika **53** (2006), 161–168.

[7] J. M. Anderson, J. G. Clunie, and Ch. Pommerenke, *On Bloch functions and normal functions*. J. Reine Angew. Math. **270** (1974), 12–37.

[8] J. Arazy, S. Fisher, and J. Peetre, *Möbius Invariant Spaces of Analytic Functions*. In: Lecture Notes in Math., Springer-Verlag, Nr. 1275 (1987), 10–22.

[9] K. Astala, J. L. Fernández, and S. Rohde, *Quasilines and the Hayman-Wu theorem*. Indiana Univ. Math. J. **42** (1993), no.4, 1077–1100.

[10] F. G. Avhadiev and L. A. Aksent'ev, *The main results on sufficient condition for an analytic function to be schlicht*. Russ. Math. Surv, **30** (1975), no.4, 1–63 (translated from Usp. Mat. Nauk, 1975, t. 30, no. 4(184), 3–63).

[11] F. G. Avkhadiev, *Solution of the Generalized Saint Venant Problem*. Sb. Math. **189** (1998), 1739–1748 (translated from Mat. Sb. **189**(12) (1998), 3–12).

[12] F. G. Avkhadiev, *Hardy Type Inequalities in Higher Dimensions with Explicit Estimate of Constants*. Lobachevskii J. Math. **21** (2006), 3–31 (electronic).

[13] F. G. Avkhadiev, *Hardy-type inequalities on planar and spatial open sets.* Proceedings of the Steklov Institute of Mathematics **255** (2006), no.1, 2–12 (translated from Trudy Matem. Inst. V.A. Steklova, 2006, v.255, 8–18).

[14] F. G. Avkhadiev and K.-J. Wirths, *Estimates of derivatives in an α-invariant set of analytic functions.* Analysis **19** (1999),19-27.

[15] F. G. Avkhadiev and K.-J. Wirths, *Convex holes produce lower bounds for coefficients.* Complex Variables **47** (2002), 553–563.

[16] F. G. Avkhadiev and K.-J. Wirths, *Schwarz-Pick inequalities for derivatives of arbitrary order.* Constr. Approx. **19** (2003), 265–277.

[17] F. G. Avkhadiev and K.-J. Wirths, *Punishing factors for angles.* Comp. Methods and Function Theory **3** (2003), 127–141.

[18] F. G. Avkhadiev and K.-J. Wirths, *Poles near the origin produce lower bounds for coefficients of meromorphic univalent functions.* Michigan Math. J. **52** (2004), 119–130.

[19] F. G. Avkhadiev and K.-J. Wirths, *Schwarz-Pick inequalities for hyperbolic domains in the extended plane.* Geom. Dedicata **106** (2004), 1–10.

[20] F. G. Avkhadiev and K.-J. Wirths, *The conformal radius as a function and its gradient image.* Isr. J. Math. **145** (2005), 349–374.

[21] F. G. Avkhadiev and K.-J. Wirths, *Punishing factors for finitely connected domains.* Monatshefte f. Math. **147** (2006), 103–115.

[22] F. G. Avkhadiev and K.-J. Wirths, *Sharp bounds for sums of coefficients of inverses of convex functions.* Comp. Methods and Function Theory **7** (2007), 105–109.

[23] F. G. Avkhadiev and K.-J. Wirths, *A proof of the Livingston conjecture.* Forum Math. **19** (2007), 149–157.

[24] F. G. Avkhadiev and K.-J. Wirths, *The punishing factors for convex pairs are 2^{n-1}.* Revista Mat. Iberoamericana **23** (2007), 847–860.

[25] F. G. Avkhadiev and K.-J. Wirths, *Punishing factors and Chua's conjecture.* Bull. Belg. Math. Soc., Simon Stevin **14** (2007), 333–340.

[26] F. G. Avkhadiev and K.-J. Wirths, *Subordination under concave univalent functions.* Complex Variables and Elliptic Equations **52** (2007), 299–305.

[27] F. G. Avkhadiev and K.-J. Wirths, *A Theorem of Teichmüller, Uniformly Perfect Sets and Punishing factors.* Preprint (Tech. Univ. Braunschweig, 2005) http://www.iaa.tu-bs.de/wirths/preprints/teich.ps

[28] F. G. Avkhadiev and K.-J. Wirths, *Estimates of the derivatives of meromorphic maps from convex domains into concave domains.* Comp. Methods and Function Theory **8** (2008), 107–119.

[29] F. G. Avkhadiev and K.-J. Wirths, *Local Taylor coefficients and subordination.* Preprint (Tech. Univ. Braunschweig, 2005) http://www.iaa.tu-bs.de/wirths/preprints/local.ps

[30] F. G. Avkhadiev, Ch. Pommerenke, and K.-J. Wirths, *On the coefficients of concave univalent functions.* Math. Nachr. **271** (2004), 3–9.

[31] F. G. Avkhadiev, Ch. Pommerenke, and K.-J. Wirths, *Sharp inequalities for the coefficients of concave schlicht functions.* Comment. Math. Helv. **81** (2006), 801–807.

[32] A. Baernstein, II, *Integral means, univalent functions and circular symmerization.* Acta Math. **133** (1974), 139–169.

[33] A. Baernstein, II and G. Schober, *Estimates for inverse coefficients of univalent functions from integral means.* Israel J. Math. **36** (1980), 75-82.

[34] C. Bandle and M. Flucher, *Harmonic radius and concentration of energy, hyperbolic radius and Liouville's equation* $\triangle U = e^U$ *and* $\triangle U = U^{\frac{n+2}{n-2}}$. SIAM Review **38** (1996), 191–238.

[35] H. Bateman, *The k-function, a particular case of the confluent hypergeometric function.* Trans. Amer. Math. Soc. **33** (1931), 817–831.

[36] A. F. Beardon and F. W. Gehring, *Schwarzian derivatives, the Poincaré metric and the kernel function.* Comment. Math. Helv. **55** (1980), 50–64.

[37] A. F. Beardon and Ch. Pommerenke, *The Poincaré metric of plane domains.* J. London Math. Soc. (2) **18** (1978), 475–483.

[38] C. Bénéteau, A. Dahlner, and D. Khavinson, *Remarks on the Bohr phenomenon.* Comp. Methods and Function Theory **4** (2004), 1–19.

[39] A. Bermant, *Dilatation d' une fonction modulaire et problèmes de recouvrement (in Russian, French summary).* Matematicheskii Sbornik **15(57)** (1944), No. 2, 285–318.

[40] L. Bieberbach, *Über die Koeffizienten derjenigen Potenzreihen, welche eine schlichte Abbildung des Einheitskreises vermitteln.* Sitz. Ber. Preuss. Akad. Wiss., (1916), 940–955.

[41] M. Biernacki, *Sur la représentation conforme des domaines linéarement accessibles.* Prace Mat. Fiz. **44** (1936), 293–314.

[42] É. Borel, *Sur les zéros des fonctions entières.* Acta Math. **20** (1896 - 97), 357-396.

[43] L. de Branges, *A proof of the Bieberbach conjecture.* Acta Math. **154** (1985), 137-152.

[44] D. A. Brannan, *On coefficient problems for certain power series.* Proc. of the Symposium on Complex Analysis (Univ. Kent, Canterbury,1973, London Mat. Soc. Lecture Note Series, Cambridge Univ. Press) **12** (1974), 17–27.

[45] D. A. Brannan, J. G. Clunie, and W. E. Kirwan, *On the coefficient problem for functions of bounded boundary rotation.* Ann. Acad. Sci. Fenn. AI **523** (1973), 18p.

[46] L. Brickman, *Subordinate families of analytic functions.* Illinois J. Math. **15** (1971), 241–248.

[47] J. E. Brown, *Iterations of functions subordinate to schlicht functions.* Complex Variables **9** (1987), 143–152.

[48] R. B. Burckel, *An Introduction to Classical Complex Analysis, Vol.1.* Birkhäuser Verlag, Basel und Stuttgart, 1979.

[49] L. A. Caffarelli and A. Friedman, *Convexity of solutions of semilinear elliptic equations.* Duke Math. J. **52** (1985), 15–44.

[50] J. T. P. Campschroer, *Growth of the coefficients of the inverse of a convex function.* Indag. Math. **46** (1984), 1–6.

[51] C. Carathéodory, *Über den Variabilitätsbereich der Koeffizienten von Potenzreihen, die gegebene Werte nicht annehmen.* Math. Ann. **64** (1907), 95–115.

[52] C. Carathéodory, *Über den Variabilitätsbereich der Fourierschen Konstanten von positiven harmonischen Funktionen.* Rend. Cir. Mat. Palermo **32** (1911), 193–217.

[53] C. Carathéodory, *Untersuchungen über die konformen Abbildungen von festen und veränderlichen Gebieten.* Mat. Ann. **52**:1 (1912), 107–144.

[54] C. Carathéodory, *Theory of functions of a complex variable.* Chelsea Publ. Comp., New York, 1960.

[55] L. Carleson and T. W. Gamelin, *Complex dynamics.* Springer, New York, 1993.

[56] K. S. Chua, *Derivatives of univalent functions and the hyperbolic metric.* Rocky Mountain J. Math. **26** (1996), 63–75.

[57] J. Clunie, *On meromorphic schlicht functions.* J. London Math. Soc. **34** (1958), 215–216.

[58] J. G. Clunie, *Inverse coefficients for convex univalent functions.* Journal d'An. Math. **36** (1979), 31–35.

[59] P. L. Duren, *Theory of H^p spaces.* Academic Press, New York, 1970.

[60] P. L. Duren, *Univalent functions.* Springer, New York, 1980.

[61] L. Fejér, *Über gewisse durch die Fouriersche und Laplacesche Reihe definierten Mittelkurven und Mittelflächen.* Palermo Rend. **38** (1914), 79–97.

[62] J. L. Fernández, *Domains with strong barrier.* Rev. Math. Iberoamericana **5** (1989), no.1-2, 47–65.

[63] J. L. Fernández and J. M. Rodríguez, *The Exponent of Convergence of Riemann Surfaces. Bass Riemann Surfaces.* Ann. Acad. Sci. Fenn., Ser. A.I: Math. **15** (1990), 165–182.

[64] C. H. FitzGerald and Ch. Pommerenke, *The de Branges theorem on univalent functions.* Trans. Amer. Math. Soc. **290** (1985) 683–690.

[65] F. P. Gardiner and N. Lakic, *Comparing Poincaré densities.* Ann. Math. **154** (2001), 115–127.

[66] J. B. Garnett, *Bounded analytic functions.* Academic Press, Inc. New York, London, 1981.

[67] J. B. Garnett and D. E. Marshall, *Harmonic measure.* Cambridge Univ. Press, Cambridge, 2005.

[68] P. Ghatage and D. Zheng, *Hyperbolic derivatives and generalized Schwarz-Pick estimates.* Proc. Amer. Math. Soc. **132** (2004), 3309–3318.

[69] G. M. Goluzin, *On majorant of subordinate analytic functions I.* Mat. Sb. **29**(71) (1951), 209–224.

[70] G. M. Goluzin, *Geometric theory of functions of a complex variable.* AMS, Providence, 1969.

[71] S. Gong, *The Bieberbach Conjecture.* AMS/IP studies in advanced mathematics **12**, AMS, Internatinal Press, 1999.

[72] A. W. Goodman, *Functions typically-real and meromorphic in the unit circle.* Trans. Amer. Math. Soc. **81** (1956), 92–105.

[73] A. W. Goodman, *Univalent functions.* Mariner Publ. Comp., Tampa, FL, 1983.

[74] B. Gustafsson, *On the convexity of a solution of Liouville's equation.* Duke Math. J. **60** (1990), 303–311.

[75] D. J. Hallenbeck and T. H. MacGregor, *Linear Problems and Convexity Techniques in Geometric Function Theory.* Pitman Press, Boston, 1984.

[76] R. Harmelin, *Hyperbolic metric, curvature of geodesics and hyperbolic discs in hyperbolic domains.* Israel J. Math. **70** (1990), 111–126.

[77] R. Harmelin and D. Minda , *Quasi-invariant domain constants.* Israel J. Math **77** (1992), 115–127.

[78] W. K. Hayman, *Multivalent functions.* Cambridge Univ. Press, Cambridge, 1958.

[79] W. K. Hayman and J. M. G. Wu, *Level sets of univalent functions.* Comment. Math. Helv. **56** (1981), no. 3, 366–403.

[80] J. A. Hempel, *The Poincaré metric on the twice punctured plane and the theorems of Landau and Schottky.* J. London Math. Soc. (2) **20**(1979), 435–445.

[81] P. Henrici, *Applied Computational Complex Analysis.* Vol.1, Wiley, New York, 1974.

[82] D. A. Herron, X. Liu, and D. Minda, *Ring domains with separating circles or separating annuli.* J. Analyse Math. **53** (1989), 233–252.

[83] A. Herzig, *Die Winkelderivierte und das Poisson-Stieltjes-Integral.* Math. Z. **46** (1940), 129–156.

[84] J. A. Hummel, St. Scheinberg, and L. Zalcman, *A coefficient problem for bounded nonvanishing functions.* J. Anal. Math. **31** (1977), 169–190.

[85] Z. J. Jakubowski, *On the upper bound of the functional $\left| f^{(n)}(z) \right|$ ($n = 2, 3, \ldots$) in some classes of univalent functions.* Comment. Math. Prace Mat. **17** (1973), 65–69.

[86] J. A. Jenkins, *On a conjecture of Goodman concerning meromorphic univalent functions.* Michigan Math. J. **9** (1962), 25-27.

[87] J. A. Jenkins, *On explicit bounds in Landau's theorem II.* Can. J. Math. **33** (1981), 559–562.

[88] V. Jørgensen, *On an inequality for the hyperbolic measure and its application in the theory of functions.* Math. Scand. **4** (1956), 113–124.

[89] S.-A. Kim and C. D. Minda, *The hyperbolic and quasihyperbolic metrics in convex regions.* J. Anal. **1** (1993), 109–118.

[90] W. E. Kirwan and G. Schober, *Inverse coefficients for functions of bounded boundary rotation.* Journal d'An. Math. **36** (1979), 167–178.

[91] R. Klouth and K.-J. Wirths, *Two new extremal properties of the Koebe-function.* Proc. Amer. Math. Soc. **80** (1980), 594–596.

[92] G. Knese, *A Schwarz lemma on the polydisk.* Proc. Amer. Math. Soc. **135** (2007), 2759–2768.

[93] W. Koepf, *On close-to-convex functions and linearly accessible domains.* Complex Var. **11** (1989), 269–279.

[94] W. Koepf and D. Schmersau, *Bounded nonvanishing functions and Bateman functions.* Compl. Var. **25** (1994), 237–259.

[95] L. V. Kovalev, *Domains with convex hyperbolic radius.* Acta Math. Univ. Comenianae **70** (2001), 207–213.

[96] J. G. Krzyż, *Problem 1, posed in : fourth conference on analytic functions.* Ann. Polon. Math. **20** (1967-1968), 314.

[97] R. Kühnau, *The conformal modul of quadrilaterals and of rings.* Handbook in complex analysis: Geometric Function Theory **2** (2005), 99–129, Elsevier, Amsterdam.

[98] E. Landau, *Einige Bemerkungen über schlichte Abbildung.* J.ber. Deutsche Math. Verein. **34** (1925/26), 239–243.

[99] N.A. Lebedev and I. M. Milin, *An inequality.* Vestnik Leningrad Univ. **20** (1965), 157–158 (in Russian).

[100] O. Lehto, *On the distortion of conformal mappings with bounded boundary rotation.* Ann. Acad. Sci. Fenn. Ser. AI Math. Phys. **124** (1952), 14p.

[101] V. Levin, *Aufgabe 163.* J.ber. Deutsche Math. Verein. **43** (1933), 133; Lösung, ibid. **44** (1934), 80–83 (solutions by W. Fenchel and E. Reissner).

[102] J.-L. Li, *Estimates for derivatives of holomorphic functions in a hyperbolic domain.* J. Math. Anal. Appl. **329** (2007), 581–591.

[103] J.-L. Li, *Schwarz-Pick inequalities for convex domains.* Kodai Math. J. **30** (2007), 252-262.

[104] R. J. Libera and E. J. Zlotkiewicz, *Early coefficients of the inverse of a regular convex function.* Proc. Amer. Math. Soc. **85** (1982), 225–230.

[105] E. Lindelöf, *Mémoire sur certaines inégalités dans la théorie des fonctions monogènes et sur quelques propriétés nouvelles de ces fonctions dans le voisinage d'un point singulier essentiel.* Acta Societatis Scientiarum Fennicae **XXXV** (1909), Nr.7 (1908), 35 pp.

[106] J. E. Littlewood, *On inequalities in the theory of functions.* Proc. London Math. Soc. **23** (1925), 481–519.

[107] J. E. Littlewood, *Lecture on the theory of functions.* Oxford University Press, Oxford, 1944.

[108] A. E. Livingston, *Convex meromorphic mappings.* Ann. Pol. Math. **59.3** (1994), 275–291.

[109] K. Löwner, *Untersuchungen über die Verzerrung bei konformen Abbildungen des Einheitskreises $|z| < 1$, die durch Funktionen mit nicht verschwindender Ableitung geliefert werden.* Leipziger Berichte **69** (1917), 89–106.

[110] K. Löwner, *Untersuchungen über schlichte konforme Abbildungen des Einheitskreises I.* Math. Ann. **89** (1923), 103–121.

[111] W. Ma and D. Minda, *Behavior of domain constants under conformal mappings.* Israel J. Math. **91** (1995), 157–171.

[112] B. MacCluer, K. Stroethoff, and R. Zhao , *Generalized Schwarz-Pick estimates.* Proc. Amer. Math. Soc. **131** (2002), 593–599.

[113] B. MacCluer, K. Stroethoff, and R. Zhao, *Schwarz-Pick estimates.* Complex Variables **48** (2003) 711–730.

[114] A. J. Macintyre and W. W. Rogosinski, *Extremum problems in the theory of analytic functions.* Acta Math. **82** (1950), 275–325 .

[115] A. Marx, *Untersuchungen über schlichte Abbildungen.* Math. Ann. **107** (1932/33), 40–65.

[116] I. M. Milin, *On the coefficients of univalent functions.* Soviet Math. Doklady **8** (1967), 1255–1258.

[117] J. Miller, *Convex and starlike meromorphic functions.* Proc. Amer. Math. Soc. **80** (1980), 607–613.

[118] S. S. Miller and P. T. Mocanu, *Differential subordination, Theory and applications.* Marcel Dekker Inc., New York, 2000.

[119] C. D. Minda and D. J. Wright, *Univalence criteria and the hyperbolic metric in convex regions.* Rocky Mtn. J. Math. **12** (1982), 471–479.

[120] R. Nevanlinna, *Eindeutige Analytische Funktionen.* Springer-Verlag, Berlin, 1953.

[121] B. G. Osgood, *Some properties of f''/f' and the Poincaré metric.* Indiana Univ. Math. J. **31** (1982), 449–461.

[122] V. Paatero, *Über die konforme Abbildung von Gebieten deren Ränder von beschränkter Drehung sind.* Ann. Acad. Sci. Fenn. Ser. A **9** (1931), 77 pp.

[123] J. Pfaltzgraff and B. Pinchuk, *A variational method for classes of meromorphic functions.* J. Analyse Math. **24** (1971), 101–150.

[124] G. Pick, *Über allgemeine Konvergenzsätze der komplexen Funktionentheorie.* Mat. Ann. **77**:1 (1916), 1–6.

[125] G. Pick, *Über die Beschränkungen analytischer Funktionen, welche durch vorgeschriebene Funktionswerte bewirkt werden.* Mat. Ann. **77** (1916), 7–23.

[126] Ch. Pommerenke, *Linear-invariante Familien analytischer Funktionen. I.* Math. Ann. **155** (1964), 108–154.

[127] Ch. Pommerenke, *On close-to-convex analytic functions.* Trans. Amer. Math. Soc. **114** (1965), 176–186.

[128] Ch. Pommerenke, *Univalent functions.* Vandenhoeck and Ruprecht, Göttingen, 1975.

[129] Ch. Pommerenke, *Uniformly perfect sets and the Poincaré metric.* Arch. Math. **32** (1979), 192–199.

[130] Ch. Pommerenke, *On uniformly perfect sets and Fuchsian groups.* Analysis **4** (1984), 299–321.

[131] Ch. Pommerenke, *Boundary behaviour of conformal maps.* Springer, New York, 1992.

[132] Ch. Pommerenke, *Personal Communication.* 3. 12. 2002.

[133] D. V. Prokhorov, *Bounded univalent functions.* Handbook in complex analysis: Geometric Function Theory **1** (2005), 207–228, North-Holland, Amsterdam.

[134] J. Radon, *Über die Randwertaufgaben beim logarithmischen Potential.* Sitzungsberichte der Oesterreichischen Akademie der Wissenschaften **128** (1920), 1123–1167.

[135] M. S. Robertson, *On the coefficients of typically-real functions.* Bull. Amer. Math. Soc. **41** (1935), 565–572.

[136] M. S. Robertson, *A remark on odd schlicht functions.* Bull. Amer. Math. Soc.**42** (1936), 366–370.

[137] M. S. Robertson, *Quasi-subordination and coefficient conjectures.* Bull. Amer. Math. Soc. **76** (1970), 1–9.

[138] W. Rogosinski, *Über positive harmonische Entwicklungen und typisch reelle Potenzreihen.* Math. Z. **35** (1932), 93–121.

[139] W. Rogosinski, *On the coefficients of subordinate functions.* Proc. London Math. Soc. **48** (1943), 48–82.

[140] O. Roth and K.-J. Wirths, *Taylor coefficients of negative powers of schlicht functions.* Comp. Methods and Function Theory **1** (2001), 521–533.

[141] St. Ruscheweyh, *Über die Faltung schlichter Funktionen.* Math. Z. **128** (1972), 85–92.

[142] St. Ruscheweyh, *Über einige Klassen in Einheitskreis holomorpher Funktionen.* Ber. Math.-Stat. Sektion Forschungszentrum Graz **7** (1974), 12p.

[143] St. Ruscheweyh, *Two remarks on bounded analytic functions.* Serdica, Bulg. math. publ. **11** (1985), 200–202.

[144] St. Ruscheweyh and T. Sheil-Small, *Hadamard products of schlicht functions and the Pólya-Schoenberg conjecture.* Comment. Math. Helv. **48** (1973), 119–135.

[145] N. Samaris, *A proof of Krzyż's conjecture for the fifth coefficient.* Complex Variables **48** (2003), 753–766.

[146] G. Schober, *Univalent Functions - Selected Topics.* Lecture Notes in Math. **478**, Springer, New York, 1975.

[147] G. Schober, *Coefficient estimates for inverses of schlicht functions.* In: Aspects of contemporary complex analysis, Academic Press, New York, 1980.

[148] I. Schur, *Über Potenzreihen, die im Innern des Einheitskreises be-schränkt sind.* J. Reine Angew. Math. **147** (1917), 205–232.

[149] H. A. Schwarz, *Gesammelte Abhandlungen.* Vol.II, Springer, Berlin, 1890.

[150] J. H. Shapiro, *Composition operators and classical function theory.* Springer, New York 1993.

[151] T. Sheil-Small, *On convex univalent functions.* J. London Math. Soc.(2)**1** (1969), 483–492.

[152] T. Sheil-Small, *On the convolution of analytic functions.* J. Reine Angew. Math. **258** (1973), 137–152.

[153] A. Yu. Solynin, *On separation of continua by circles* (Russian). Zap. Nauchn. Semin. Leningr. Otd. Mat. Inst. Steklova **168** (1988), 154–157.

[154] A. Yu. Solynin and M. Vuorinen, *Estimates for the hyperbolic metric of the punctured plane and applications.* Isr. J. Math. **124** (2001), 29–60.

[155] E. Strohhäcker, *Beiträge zur Theorie der schlichten Funktionen.* Math. Z. **37** (1933), 356–380.

[156] T. J. Suffridge, *Some remarks on convex maps of the unit disc.* Duke Math. J. **37** (1970), 775–777.

[157] T. J. Suffridge, *Problems for Non-vanishing H^p Functions.* In: Lecture Notes in Math.(1990) Nr. 1435, 177–190.

[158] W. Szapiel, *A new approach to the Krzyż conjecture.* Ann. Univ. Mariae Curie-Sklodowska Sect. A **48** (1994), 169–192.

[159] O. Szász, *Ungleichungen für die Koeffizienten einer Potenzreihe.* Math. Z. **1** (1918), 163–183.

[160] O. Szász, *Ungleicheitsbeziehungen für die Ableitungen einer Potenzreihe, die eine im Einheitskreis beschränkte Funktion darstellt.* Math. Z. **8** (1920), 303–309.

[161] D. L. Tan, *Estimates of coefficients of bounded nonvanishing analytic functions.* (Chinese), Chinese Ann. Math., Ser. A **4** (1983), 97–104. English summary in Chinese Ann. Math., Ser. B **4** (1983), 131-132.

[162] O. Teichmüller, *Untersuchungen über konforme und quasikonforme Abbildungen.* Deutsche Math. **3** (1938), 621–678.

[163] P. Todorov, *New explicit formulas for the nth derivative of composite functions.* Pacific J. Math. **92** (1981), 217–236.

[164] S. Y. Trimble, *A coefficient inequality for convex univalent functions*. Proc. Amer. Math. Soc. **48** (1975), 266–267.

[165] K.-J. Wirths, *Über holomorphe Funktionen, die einer Wachstumsbeschränkung unterliegen*. Arch. Math. **30** (1978), 606–612.

[166] K.-J. Wirths, *The Koebe domain for concave univalent functions*. Serdica Math. J. **29** (2003), 355-360.

[167] K.-J. Wirths, *On the residuum of concave univalent functions*. Serdica Math. J. **32** (2006), 209–214.

[168] S. Yamashita, *La dérivée d'une fonction univalente dans une domaine hyperbolique*. C. R. Acad. Sci. Paris Ser I. Math. **314** (1992), 45–48.

[169] S. Yamashita, *Localization of the coefficient Theorem*. Kodai Math. J. **22** (1999), 384–401.

[170] S. Yamashita, *Higher derivatives of holomorphic functions with positive real part*. Hokkaido Math. J. **29** (2000), 23-36.

Index